Animal Ribosomes: Experimental Studies Of The Last Five Years

Papers by
Giuseppe Attardi, William Steele, Kenneth Tartof,
Robert Perry, et al.

MSS Information Corporation
19 East 48th Street New York, N.Y. 10017

TABLE OF CONTENTS

CREDITS AND ACKNOWLEDGEMENTS

Amaldi, Francesco; and Giuseppe Attardi, "A Comparison of the Primary Structures of 28S and 18S Ribonucleic Acid from HeLa Cells and Different Human Tissues," *Biochemistry*, 1971, 10:1478-1483.

Attardi, Barbara; Barbara Cravioto; and Giuseppe Attardi, "Membrane-Bound Ribosomes in HeLa Cells: I. Their Proportion to Total Cell Ribosomes and Their Association with Messenger RNA," *Journal of Molecular Biology*, 1969, 44:47-70.

Blobel, G.; and Van R. Potter, "Ribosomes in Rat Liver: An Estimate of the Percentage of Free and Membrane-Bound Ribosomes Interacting with Messenger RNA *in vivo*," *Journal of Molecular Biology*, 1967, 28:539-542.

Cocucci, S. M.; and M. Sussman, "RNA in Cytoplasmic and Nuclear Fractions of Cellular Slime Mold Amebas," *The Journal of Cell Biology*, 1970, 45:399-407.

Grossbach, Ulrich; and I. Bernard Weinstein, "Separation of Ribonucleic Acids by Polyacrylamide Gel Electrophoresis," *Analytical Biochemistry*, 1968, 22:311-320.

Jeanteur, Ph.; and G. Attardi, "Relationship Between HeLa Cell Ribosomal RNA and Its Precursors Studied by High Resolution RNA-DNA Hybridization," *Journal of Molecular Biology*, 1969, 45:305-324.

Mohan, Jag; and F. M. Ritossa, "Regulation of Ribosomal RNA Synthesis and Its Bearing on the Bobbed Phenotype in *Drosophila melanogaster*," *Developmental Biology*, July, 1970, 22:495-512.

Pardue, Mary Lou; Susan A. Gerbi; Ronald A. Eckhardt; and Joseph G. Gall, "Cytological Localization of DNA Complementary to Ribosomal RNA in Polytene Chromosomes of *Diptera*," *Chromosoma* (Berl.), 1970, 29:268-290.

Ritossa, F. M., "Unstable Redundancy of Genes for Ribosomal RNA," *Proceedings of the National Academy of Sciences*, June, 1968, 60:509-516.

Shimizu, Nobuyoshi; and Kin-Ichiro Miura, "Studies on Nucleic Acids of Living Fossils: I. Isolation and Characterization of DNA and Some RNA Components from the Brachiopod Lingula," *Biochimica et Biophysica Acta*, 1971, 232:271-277.

Steele, William J., "Localization of Deoxyribonucleic Acid Complementary to Ribosomal Ribonucleic Acid and Preribosomal Ribonucleic Acid in the Nucleolus of Rat Liver," *The Journal of Biological Chemistry*, June 25, 1958, 243:3333-3341.

Tartof, Kenneth D., "Increasing the Multiplicity of Ribosomal RNA Genes in *Drosophila melanogaster*," *Science*, January 22, 1971, 294-297.

Tartof, Kenneth D.; and Robert P. Perry, "The 5 d RNA Genes of *Drosophila melanogaster*," *Journal of Molecular Biology*, 1970, 51:171-183.

Weinberg, Robert A.; Ulrich Loening; Margherita Willems; and Sheldon Penman, "Acrylamide Gel Electrophoresis of HeLa Cell Nucleolar RNA," *Proceedings of the National Academy of Sciences*, September, 1967, 58:1088-1095.

Weinberg, Robert A.; and Sheldon Penman, "Processing of 45 s Nucleolar RNA," *Journal of Molecular Biology*, 1970, 47:169-178.

A Comparison of the Primary Structures of 28S and 18S Ribonucleic Acid from HeLa Cells and Different Human Tissues

Francesco Amaldi and Giuseppe Attardi

A redundancy in the genetic information for the two high molecular weight rRNA components and for 5S rRNA has been demonstrated by RNA–DNA hybridization in bacteria and, of two or more orders of magnitude greater, in eukaryotic cells (see review by Attardi and Amaldi, 1970). In HeLa cells, a line of human origin, the number of genes has been found to be about 1000/cell for each of the two high molecular weight components, 28S and 18S RNA (Jeanteur and Attardi, 1969), and about 7600 for the 5S RNA (Hatlen and Attardi, 1971). This redundancy of information raises the problem of the possible variability of the rRNA genes.

The existence of such variability in a higher organism could result either in heterogeneity of the populations of each of the rRNA components within the same cell, or, if all rRNA genes are not equally expressed in different cell types of the same organism, in differences between rRNA preparations from different tissues or developmental stages of the same organism.

Reich et al. (1963) had reported substantial differences in base composition between samples of unfractionated rRNA from different tissues of the same animal species, more so than between RNA samples prepared from the same tissues of different animal species. These results, however, were not confirmed by Hirsch (1966), who found the same base composition in 28S RNA, and, respectively, in 18S RNA, purified from various rat and rabbit tissues. Likewise, other investigators could not detect any difference in base composition between preparations of the two high molecular weight rRNA components or of unfractionated rRNA isolated from different developmental stages of the same organism (Lerner et al., 1963; Henney and Storck, 1963; Brown and Gurdon, 1964; Slater and Spiegelman, 1966; Tata, 1967; Grummt and Bielka, 1968). Gould et al. (1966) analyzed by polyacrylamide gel electrophoresis the products of limited digestion by T1 ribonuclease of rRNA (unfractionated) prepared from different organisms and from two cell types, reticulocytes and liver cells, of the same organism (rabbit): while they found differences in the pattern of degradation of the rRNA from different organisms, the rRNA from the two rabbit cell types analyzed gave identical results. Similarly, RNA–DNA hybridization experiments failed to show any difference in sequence between high molecular weight rRNA components prepared from rabbit reticulocytes and liver (Di Girolamo et al., 1969), or from different developmental stages of sea urchin (Mutolo and Giudice, 1967).

The above-mentioned negative results could conceivably reflect the insensitivity of the analytical methods employed. In order to approach the problem at a finer level of analysis, in the present work, the method of oligonucleotide mapping previously described (Amaldi and Attardi, 1968) has been used to compare the distribution of nucleotide sequences released by pancreatic RNase in 28S and 18S RNA prepared from HeLa cells and from different human tissues. The results obtained by this method as well as by other methods (base composition studies, sedimentation analysis after limited RNase digestion, and RNA–DNA hybridization) indicate that, even at the highest level of resolution employed here, there are no detectable differences between the 28S RNA and, likewise, 18S RNA preparations analyzed.

Materials and Methods

Cells and Tissues. Reference is made to a preceding paper (Amaldi and Attardi, 1968) for the method of growth of HeLa cells.

Human tissues were mainly autoptic material, and in a few cases surgical material, derived from male and female individuals (from 6 months to 67 years in age). The tissues were frozen as soon as possible and kept at $-70°$ until used for RNA extraction.

Labeling Conditions. The method utilized to label HeLa cell rRNA to a high specific activity with [^{32}P]orthophosphate has been described previously (Attardi *et al.*, 1965b).

RNA Extraction. ^{32}P-labeled 28S and 18S RNA were extracted from the purified 50S and 30S ribosomal subunits, respectively; nonradioactive HeLa rRNA was isolated from total cells; rRNA of human tissues, on the contrary, was extracted from the monomer–polysome fraction. The methods of isolation of ribosomal subunits, extraction and fractionation of rRNA into the 28S and 18S components have been previously described (Amaldi and Attardi, 1968).

Base Composition and Sequence Analysis. The procedure utilized for base composition analysis and oligonucleotide mapping after pancreatic RNase digestion have been reported in detail in a preceding paper (Amaldi and Attardi, 1968).

Hybridization Experiments. RNA–DNA hybridization and isolation of the hybrids have been carried out by RNase digestion and Sephadex chromatography as described previously (Attardi *et al.*, 1965a). In some experiments, the RNA–DNA hybrids were collected directly on Millipore membranes after RNase digestion (30 μg/ml, 30 min at 37°) and washed with 100 ml of 0.5 M KCl–0.01 M Tris buffer (pH 7.0) at 60°.

Partial Degradation of RNA by Mild Digestion with Pancreatic or T1 Ribonuclease. The nonradioactive RNA to be analyzed (about 100 μg) was mixed with a small amount of ^{32}P-labeled HeLa 28S RNA in a volume of 0.4 ml of SSC (SSC: 0.15 M NaCl–0.015 M sodium citrate). Treatment with pancreatic RNase (five-times crystallized, Sigma Chemical Co., heated at 80° for 15 min) was carried out at 2° for 20 min with 0.005 μg of enzyme; in some experiments, the limited digestion was carried out with 1 unit of T1 RNase (Sankyo Co., Ltd., Tokyo). At the end of the digestion, sodium dodecyl sulfate was added to a final concentration of 0.5%, and the sample was layered on a 5–20% sucrose gradient in SSC and centrifuged in the SW25.3 rotor of the Spinco L ultracentrifuge at 25,000 rpm for 20 hr at 2°. The collected fractions were analyzed for ultraviolet absorption and for ^{32}P acid-precipitable radioactivity, as previously described (Attardi et al., 1966).

Results

Isolation of 28S and 18S rRNA from HeLa Cells and from Human Tissues. The isolation and the criteria of purity of 28S and 18S RNA (^{32}P labeled and nonlabeled) from HeLa cells have been discussed previously (Amaldi and Attardi, 1968).

The RNA of human tissues was mainly prepared from autoptic material; for this reason it was expected that in many cases it might have undergone some degradation. As criteria to distinguish intact from slightly degraded RNA preparations a ratio of 28S to 18S RNA in the first sucrose gradient centrifugation of about 2.5 and the lack of material between the 18S and the 4–5S peaks were used (Amaldi and Attardi, 1968). Only those tissue RNA preparations which satisfied these criteria have been utilized in the present work. As an example, Figure 1 shows the sedimentation pattern of an undegraded rRNA preparation from human liver.

Base Composition. Table I summarizes the results obtained for the base composition of the two rRNA components extracted from different human tissues and from HeLa cells. It appears that the base compositions of all the 28S RNA and, respectively, all the 18S RNA preparation analyzed are, within experimental error, identical.

The amount of pseudouridylic acid in the rRNA from human tissues has not been precisely determined, but it appears to be present in about the same proportion as in HeLa cells (Amaldi and Attardi, 1968).

Partial Sequence Analysis. The analysis of frequencies of mono-, di-, and trinucleotides released by pancreatic RNase digestion has been carried out as described in detail in a preceding paper (Amaldi and Attardi, 1968). The method of the relative specific activity has been utilized here. In brief, this method consists in carrying out the RNase digestion and DEAE chromatography on a mixture of a relatively large amount of an unlabeled RNA species and a small amount of a ^{32}P-labeled RNA species and in determining the specific activity of the individual RNase digestion products (up to trinucleotides), and then normalizing it to the average specific activity (thus obtaining the relative specific activity). By using either the unlabeled or labeled RNA as an internal standard, one can compare different labeled or, respectively, unlabeled RNA samples. Although this method does not give absolute values but only relative values for the frequencies of the products of RNase digestion, it is very accurate and can reveal small differences between various RNA preparations.

^{32}P-labeled HeLa 28S RNA has been used in the present work as an internal standard to compare the nonradioactive preparations of 28S RNA (and 18S RNA) from HeLa cells and from different human tissues. Table II summarizes the results obtained by this analysis. In the table the results are expressed as ratios between the relative specific activities obtained in experiments utilizing ^{32}P 28S RNA from HeLa cells and nonradioactive 28S RNA (or 18S RNA) from different human tissues and the relative specific activities obtained in experiments utilizing ^{32}P 28S RNA both and

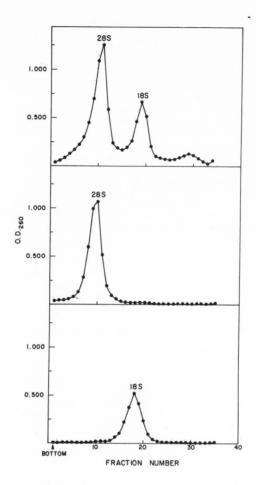

FIGURE 1: Sedimentation pattern of rRNA extracted from the monomer–polysome fraction of human liver. The RNA was extracted from this fraction with sodium dodecyl sulfate–phenol (Attardi *et al.*, 1965b) and separated into 28S and 18S components by two cycles of centrifugation in 5–20% (w/w) sucrose gradient in 0.01 M acetate buffer (pH 5.0)–0.1 M NaCl (16 hr at 23,000 rpm, 2°, in the SW 25-1 Spinco rotor).

nonradioactive 28S RNA (or 18S RNA) from HeLa cells. The values thus obtained give the frequencies of the oligonucleotide sequences in the HeLa cell RNA relative to those found in the tissue RNA. It appears that the frequencies of the oligonucleotides (up to trinucleotides) released by pancreatic RNase digestion are very similar in all the 28S RNA preparations analyzed and, likewise, in all the 18S RNA preparations. The data for 18S RNA show a somewhat higher variability than those for 28S RNA, but always within the experimental error. This is probably due to the difficulty to obtain a completely undegraded rRNA from human tissues; as a consequence the 18S RNA from this source is slightly contaminated by fragments of 28S RNA (to a different extent in different preparations).

Hybridization Experiments. As a different approach to the problem of the relationship between rRNA from different cell types of the same organism, RNA–DNA hybridization experiments were performed between HeLa cell DNA and ^{32}P-labeled HeLa 28S RNA (or 18S RNA) in the presence of an excess of nonradioactive 28S RNA (or 18S RNA) from HeLa cells (as a control) or human tissues.

Figure 2 shows one of these experiments utilizing 28S RNA. It appears that the 28S RNA from all the tissues analyzed competes with [32]P HeLa 28S RNA for hybridization with the rRNA sites in the DNA to the same extent as unlabeled HeLa 28S RNA.

Similar results have been obtained in hybridization competi-

TABLE I: Nucleotide Composition of 28S and 18S RNA from HeLa Cells and Different Human Tissues.[a]

RNA Source	Moles %				No. of Determinations
	Cp	Ap	Up + ψp	Gp	
28S RNA					
HeLa cells	32.1	15.8	17.0 (1.1)	35.1	2
Human liver	32.3	15.6	16.4	35.7	2
Human spleen	32.2	15.9	16.7	35.2	2
Human kidney	32.6	15.6	16.9	34.9	2
Human pancreas	32.6	15.6	16.6	35.2	2
Human brain	30.8	16.6	17.0	35.6	1
18S RNA					
HeLa cells	27.8	19.9	21.3 (1.5)	31.0	5
Human liver	28.0	19.9	21.6	30.5	2
Human spleen	28.5	20.5	21.5	30.5	2
Human kidney	27.9	19.7	21.4	31.0	2
Human pancreas	28.1	19.8	21.5	30.6	2
Human brain	28.2	19.9	21.5	30.4	1

[a] The nucleotide composition of the unlabeled rRNA components from HeLa cells and different human tissues was determined from the optical density measurements on the basis of the extinction coefficients of the four 2′,3′-nucleotides (The extinction coefficients reported by Beaven et al. (1955) were utilized here, after correction for the different pH of the individual elution media.)

11

TABLE II: Ratios of Frequencies of Partial Nucleotide Sequences in 28S, and, Respectively, 18S RNA, from HeLa Cells and Different Human Tissues.[a]

28S RNA

RNase Digestion Product	Liver (9)[b]	Spleen (4)	Kidney (5)	Pancreas (4)
Cp	0.998 ± 0.017	0.990 ± 0.021	1.006 ± 0.013	0.998 ± 0.015
Up	1.009 ± 0.015	1.006 ± 0.027	1.026 ± 0.026	1.019 ± 0.020
ApCp	0.998 ± 0.020	0.993 ± 0.016	0.990 ± 0.021	0.995 ± 0.013
ApUp	1.044 ± 0.038	1.041 ± 0.042	1.053 ± 0.036	1.043 ± 0.042
GpCp	1.001 ± 0.013	0.994 ± 0.013	0.989 ± 0.019	0.983 ± 0.021
GpUp	1.016 ± 0.017	1.026 ± 0.039	1.007 ± 0.023	1.010 ± 0.019
ApApCp	1.013 ± 0.032	1.025 ± 0.033	0.991 ± 0.025	1.010 ± 0.028
ApApUp	1.046 ± 0.038	1.044 ± 0.045	0.996 ± 0.050	1.043 ± 0.062
GpApCp + ApGpCp	0.986 ± 0.018	0.984 ± 0.033	0.968 ± 0.019	0.976 ± 0.029
GpApUp	1.008 ± 0.032	1.017 ± 0.044	0.984 ± 0.023	1.008 ± 0.038
ApGpUp	0.983 ± 0.030	0.984 ± 0.042	0.963 ± 0.037	0.965 ± 0.039
GpGpCp	0.982 ± 0.018	1.008 ± 0.012	1.016 ± 0.017	1.018 ± 0.019
GpGpUp	1.000 ± 0.016	1.015 ± 0.031	0.998 ± 0.018	1.011 ± 0.011

18S RNA

RNase Digestion Product	Liver (4)	Spleen (4)	Kidney (4)	Pancreas (3)
Cp	0.989 ± 0.016	0.989 ± 0.015	0.985 ± 0.016	0.979 ± 0.025
Up	1.037 ± 0.044	1.013 ± 0.023	1.032 ± 0.053	1.035 ± 0.056
ApCp	0.982 ± 0.018	0.971 ± 0.032	0.984 ± 0.042	0.963 ± 0.044
ApUp	1.071 ± 0.016	1.039 ± 0.059	1.081 ± 0.105	1.107 ± 0.056
GpCp	0.972 ± 0.031	0.979 ± 0.071	0.970 ± 0.035	0.960 ± 0.044
GpUp	1.032 ± 0.033	1.019 ± 0.043	1.020 ± 0.033	1.030 ± 0.065
ApApCp	1.069 ± 0.067	1.062 ± 0.082	1.089 ± 0.093	1.059 ± 0.053
ApApUp	1.024 ± 0.064	1.013 ± 0.083	1.056 ± 0.088	1.069 ± 0.127
GpApCp + ApGpCp	1.031 ± 0.043	1.035 ± 0.036	1.036 ± 0.048	1.046 ± 0.050
GpApUp	1.032 ± 0.086	1.068 ± 0.140	1.068 ± 0.072	1.099 ± 0.084
ApGpUp	1.044 ± 0.110	0.934 ± 0.068	0.995 ± 0.056	1.027 ± 0.067
GpGpCp	0.988 ± 0.075	1.039 ± 0.077	0.990 ± 0.049	0.989 ± 0.029
GpGpUp	1.078 ± 0.026	1.064 ± 0.031	1.061 ± 0.065	1.043 ± 0.049

[a] The data summarized in this table were derived from various experiments involving the use of ^{32}P-labeled HeLa cell 28S RNA prepared from the 50S subunits and unlabeled 28S or 18S RNA from HeLa cells and from different human tissues, isolated, respectively, from total cells and from the monomer–polysome fraction.   The figures reported here represent the average ratios between relative specific activities (rsa) of pancreatic RNase digestion products obtained in experiments utilizing HeLa ^{32}P 28S RNA and unlabeled 28S RNA (or 18S RNA) from different human tissues and the relative specific activities obtained in experiments utilizing HeLa ^{32}P 28S RNA and unlabeled HeLa 28S RNA (or 18S RNA): as explained in the test,   each ratio represents the ratio of the frequencies of a given nucleotide sequence in HeLa cell RNA and tissue RNA. . (See a previous paper (Amaldi and Attardi, 1968) for the procedure utilized for the calculation of relative specific activities.) Each mean is given with its standard error multiplied by a factor corresponding to a probability of 95%,—as determined from the t distribution (Fisher, 1950). The standard error of the ratios was calculated according to the formula: $S_{\bar{x}/\bar{y}} = [(\bar{x}^2(S_{\bar{y}})^2 + \bar{y}^2(S_{\bar{x}})^2)/\bar{y}^4]^{1/2}$, where $S_{\bar{x}}$ is the standard error of $\bar{x}$ and $S_{\bar{y}}$ that of $\bar{y}$. [b] The number in parentheses   indicates the number of tissue preparations tested;   the averages of 8 rsa values for HeLa ^{32}P-28S-HeLa unlabeled 28S mixtures and 5 values for HeLa ^{32}P 28S-HeLa-unlabeled 18S mixtures were used in the calculation of the ratios in the table.

12

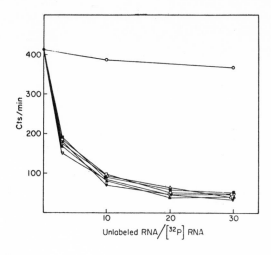

FIGURE 2: Tests of competition for sites in HeLa DNA between
³²P-labeled HeLa cell 28S RNA and unlabeled 28S RNA from HeLa
cells and different human tissues. Each incubation mixture contained, in a volume of 3 ml of 2X SSC, 50 μg of HeLa DNA, 1.0
μg of ³²P-labeled HeLa 28S RNA, and variable amounts of unlabeled 28S RNA from different source. The incubation was carried out at 70° for 4 hr and the RNA–DNA hybrids collected on
Millipore membranes after RNase treatment and washed as described in Materials and Methods. (O) *E. coli* 23S RNA; (□) Hela 28S
RNA; (■) human liver 28S RNA; (△) human spleen 28S RNA; (▲)
human kidney 28S RNA; (▽) human pancreas 28S RNA; (▼) human brain 28S RNA.

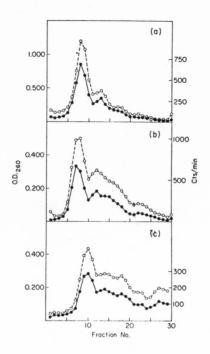

FIGURE 3: Sedimentation analysis
in sucrose gradient of 28S RNA
from HeLa cells (a), human liver
(b), and human spleen (c) after
mild digestion with pancreatic
RNase. A mixture of nonradioactive RNA from different source
(about 100 μg) and ^{32}P-labeled
HeLa 28S RNA (<1 μg) was digested with pancreatic RNase
and analyzed in sucrose gradient
as described in Materials and
Methods. (O---O) OD$_{260}$ and
(●—●) cpm of ^{32}P.

tion experiments carried out with ^{32}P HeLa 18S RNA and an excess of nonradioactive 18S RNA from HeLa cells or different human tissues.

RNA Degradation by Mild RNase Digestion. Another approach has been attempted to reveal small differences between 28S RNA preparations from various human sources. It was reasoned that minor differences in nucleotide sequences of these RNA molecules could give rise to significant differences in secondary structure, such as to be revealed by changes in sensitivity to RNase. For this reason, the RNA samples were treated, under very mild conditions, with pancreatic or T1 RNase, and the partially degraded material was analyzed by sucrose gradient centrifugation. A similar approach, involving the analysis of the electrophoretic pattern of the resistant fragments after limited T1 RNase digestion, has been utilized to compare rRNA samples from various sources (Gould *et al.*, 1966; Pinder *et al.*, 1969). In the present work, all the experiments have been performed by mixing, before the enzyme treatment, a small amount of ^{32}P HeLa 28S RNA with the nonradioactive RNA under study. The ^{32}P-labeled internal standard was necessary as a control for variations in the degree of RNase digestion in different experiments. The 28S RNA preparations from various human tissues were compared by this method with 28S RNA from HeLa cells and with each other. In most cases, no obvious differences in susceptibility to RNase were found among human liver, spleen, pancreas, kidney and HeLa 28S RNA preparations (Figure 3). Some preparations of human liver 28S RNA showed reproducibly a somewhat higher resistance to the RNase digestion than HeLa 28S; the reason for this behavior is not known and has not been investigated further.

Discussion

The purpose of this work has been to compare the primary structures of 28S and, respectively, 18S RNA preparations obtained from different human tissues. Critical to the results of this analysis was the purity of the RNA preparations utilized and the intactness of the molecules (such as to exclude the loss of any piece caused by degradation). The purity of the ^{32}P-labeled 18S and 28S RNA components, as isolated in the present work, has been discussed previously (Amaldi and Attardi, 1968). As to the unlabeled HeLa cell rRNA species, which were purified from total RNA, the only significant source of contamination was presumably the heterogeneous nuclear RNA and cytoplasmic mRNA sedimenting in the regions of 28S and 18S RNA, and the mitochondrial 16S RNA (Attardi *et al.*, 1970). Since the heterogeneous nuclear RNA represents *in toto* about 2% of total cell RNA, and the mRNA about 3% (Soeiro *et al.*, 1968), it can be estimated that the possible fraction of these heterogeneous RNA species cosedimenting with either 28S or 18S RNA would correspond to less than 2% of each rRNA component; this level of contamination would not affect appreciably the results obtained here. Somewhat more significant, although still low, would be the possible contamination of the 18S RNA by the 16S mitochondrial RNA (estimated to be at most 5%). The unlabeled rRNA species from human tissues, which were isolated from the monomer–polysome fraction, were presumably not appreciably contaminated by either heterogeneous nuclear RNA or

15

mitochondrial RNA.

As concerns the requirement of intactness of the rRNA molecules for a meaningful analysis of primary structure, in the present work it has been reasonably well satisfied by utilizing only RNA preparations which showed a normal sedimentation pattern with the expected ratio of 28S to 18S RNA (about 2.5) and without any indication of slower sedimenting degradation products.

The comparison among the various 28S RNA and, respectively, 18S RNA preparations has been carried out with various methods at different levels of resolution: (1) base composition; (2) analysis of the frequencies of mono-, di-, and trinucleotides released by pancreatic RNase digestion; (3) RNA–DNA hybridization; and (4) pattern of degradation by mild digestion with pancreatic or T1 RNase.

The first three methods proved unable to discriminate between 28S RNA and, respectively, 18S RNA samples from various sources. The fourth method, utilized only with the 28S RNA, gave similar results in most cases. A few liver 28S RNA preparations showed a higher resistance to the RNase digestion: however, the lack of reproducibility makes the interpretation of these results very difficult.

From the data presented in this work it can be concluded that all the 28S and, likewise, 18S RNA preparations analyzed are very similar, if not identical. This conclusion, which reinforces the available evidence from other eukaryotic systems, can be interpreted in different ways. (1) The genes coding for rRNA in each cell are heterogeneous, and differentially expressed in various cell types or stages of development, so that the RNA preparations studied here are indeed different from each other; however, the differences are smaller than the resolving power of our methods of analysis; (2) the rRNA genes are heterogeneous, but they are equally expressed in different cell types; as a consequence, equally heterogeneous populations of molecules would have been analyzed in this work; and (3) there is in the cell a special mechanism to keep all the rRNA genes homogeneous or nearly so during evolution.

Very little is known about the degree of homogeneity of the high molecular weight rRNA population within a given bacterial or eukaryotic cell. The available evidence points to homogeneity in sedimentation and chromatographic properties. Furthermore, homogeneity in primary structure of the high molecular weight rRNA has been suggested, in *Escherichia coli*, by the molar yield of unity or a multiple thereof of methylated sequences (Fellner and Sanger, 1968), and, in rabbit reticulocytes, by the near to unity molar yield of RNA fragments obtained by mild digestion with T1 RNase (Gould, 1967). A limited degree of heterogeneity in primary structure has, on the contrary, been shown to exist in 5S RNA from *E. coli* and strongly hinted at in 5S RNA from human source (Hatlen *et al.*, 1969).

As concerns the question of the possible heterogeneity of the rRNA genes, it should be mentioned that the rapid initial rate with which rRNA hybridizes with rDNA (Birnstiel *et al.*, 1968), the apparently regular and complete hydrogen bonding of rRNA molecules to rDNA sites in these hybrids (Jeanteur and Attardi, 1969), and the renaturation kinetics of rDNA (Birnstiel *et al.*, 1969) suggest a close similarity in sequence, if not an identity of rRNA genes. As to the mechanism which

16

may operate in keeping low the variability of the highly re-dundant rRNA genes in eukaryotic cells, one may think of a mechanism like that proposed by Callan (1967), involving one "master" gene and multiple "slave" copies, which are matched against the master and corrected for mistakes due to mutation or recombination once per life cycle of the organism or even per cell division. Alternatively, one can postulate that game-les represent a "bottleneck," where replication of a new family of rRNA genes proceeds from a master gene with destruction of the old replicas. Experimental evidence speaks, on the contrary, against a selective replication of a master gene at each cell division (Amaldi et al., 1969).

Acknowledgments

We thank Mrs. L. Wenzel and Mrs. B. Keeley for their excellent help.

References

Amaldi, F., and Attardi, G. (1968), *J. Mol. Biol. 33*, 737.

Amaldi, F., Zito-Bignami, R., and Giacomoni, E. (1969), *Eur. J. Biochem. 11*, 419.

Attardi, G., Aloni, Y., Attardi, B., Ojala, D., Pica-Mattoccia, L., Robberson, D., and Storrie, B. (1970), *Cold Spring Harb. Symp. Quant. Biol. 35*, 599.

Attardi, G., and Amaldi, F. (1970), *Annu. Rev. Biochem. 39*, 183.

Attardi, G., Huang, P. C., and Kabat, S. (1965a), *Proc. Nat. Acad. Sci. U. S. 53*, 1490.

Attardi, G., Huang, P. C., and Kabat, S. (1965b), *Proc. Nat. Acad. Sci. U. S. 54*, 185.

Attardi, G., Parnas, H., Hwang, M.-I. H., and Attardi, B. (1966), *J. Mol. Biol. 20*, 145.

Beaven, G. H., Holiday, E. R., and Johnson, E. A. (1955), *in* The Nucleic Acids, Vol. 1, Chargaff, E., and Davidson, J. N., Ed., New York, N. Y., Academic Press, p 493.

Birnstiel, M., Grunstein, M., Speirs, J., and Hennig, W. (1969), *Nature (London) 223*, 1265.

Birnstiel, M., Speirs, J., Purdom, I., Jones, K., Loening, M. E. (1968), *Nature (London) 219*, 454.

Brown, D. D., and Gurdon, J. B. (1964), *Proc. Nat. Acad. Sci. U. S. 51*, 139.

Callan, H. G. (1967), *J. Cell Sci. 2*, 1.

Di Girolamo, A., Busiello, E., and Di Girolamo, M. (1969), *Biochim. Biophys. Acta 182*, 169.

Fellner, P., and Sanger, F. (1968), *Nature (London) 219*, 236.

Fisher, R. A. (1950), Statistical Methods for Research Work-ers, Edinburgh, Oliver & Boyd.

Gould, H. J. (1967), *J. Mol. Biol. 29*, 307.

Gould, H. J., Bonanou, S., and Kanagalingham, K. (1966), *J. Mol. Biol. 22*, 369.

Grummt, F., and Bielka, H. (1968), *Biochim. Biophys. Acta 161*, 253.

Hatlen, L. E., Amaldi, F., and Attardi, G. (1969), *Biochemis-try 8*, 4989.

Hatlen, L. E., and Attardi, G. (1971), *J. Mol. Biol.* (in press).

Henney, H. R., and Storck, R. (1963), *Science 142*, 1675.

Hirsch, C. A. (1966), *Biochim. Biophys. Acta 123*, 246.

Jeanteur, Ph., and Attardi, G. (1969), *J. Mol. Biol. 45*, 305.

17

Lerner, A. M., Bell, E., and Darnell, J. E. (1963), *Science* *141*, 1187.

Mutolo, V., and Giudice, G. (1967), *Biochim. Biophys. Acta* *149*, 291.

Pinder, J. C., Gould, H. J., and Smith, I. (1969), *J. Mol. Biol.* *40*, 289.

Reich, E., Acs, G., Mach, B., and Tatum, E. L. (1963), *in* Informational Macromolecules, Vogel, H. J., Bryson, V., and Lampen, J. O., Ed., New York, N. Y., Academic Press, p 317.

Slater, D. W., and Spiegelman, S. (1966), *Biophys. J. 6*, 385.

Soeiro, R., Vaughan, M., Warner, J. R., and Darnell, J. E. (1968), *J. Cell Biol. 39*, 112.

Tata, J. R. (1967), *Biochem. J. 105*, 783.

STUDIES ON NUCLEIC ACIDS OF LIVING FOSSILS

I. ISOLATION AND CHARACTERIZATION OF DNA AND SOME RNA COMPONENTS FROM THE BRACHIOPOD LINGULA

NOBUYOSHI SHIMIZU AND KIN-ICHIRO MIURA

INTRODUCTION

The rates of evolution have not always been uniform within a species or among all organisms. In some species, change has been extremely slow for long periods of time. The Brachiopod Lingula is a rare example of this category of "living fossils". The ancestor of this marine animal is present in the earliest Cambrian deposits, and perhaps a single genus, Lingula, has persisted from the Ordovician to the present, a span of $4 \cdot 10^8$ years. It may be the oldest genus in existence[1]. In view of its slow rate of evolution and unique position with regard to animal phylogeny[2], our interest was to examine the genetic materials of Lingula to see whether it would exhibit special characteristics. The present paper describes the characterization of DNA and some species of RNA molecules from the Brachiopod Lingula.

MATERIALS AND METHODS

The following commercial products were used: DEAE-cellulose (0.93 mequiv., Brown); Sephadex G-100, G-200 and Dextran blue 2000 (Pharmacia); AG-1 × 2 (Bio-Rad); ribonuclease A, pancreatic deoxyribonuclease and snake venom phosphodiesterase (Worthington); ribonuclease T_1 and T_2 (Sankyo); protease (type VI, Sigma); radioactive amino acids (The Radiochemical Center). The aminoacyl-tRNA synthetase (EC 6.1.1 group) partially purified from baker's yeast was kindly supplied by Mr. M. Kawata.

19

Preparation of RNA. Frozen tissues (the contents of a shell) of *Lingula ana-tina*[3] were homogenized with 2 volumes of solution A (0.25 M sucrose, 5 mM MgCl$_2$, 25 mM KCl, 1 % bentonite and 50 mM Tris–HCl, pH 7.5) in a mixer at 4°. The homogenate was centrifuged at 8000 $\times$ g for 20 min and the supernatant again at 105 000 $\times$ g for 2 h. RNA was extracted from the supernatant by the phenol method[4]. Approx. 15 mg of RNA (105 000 $\times$ g-supernatant RNA) were obtained from 100 g (wet weight) of tissues. For mass preparation, frozen tissues were homogenized with 2 volumes of 45 % phenol in Tris–Mg^{2+} buffer (5 mM MgCl$_2$ and 10 mM Tris–HCl, pH 7.6) in a mixer at 4° and then stirred gently for 45 min. After settling overnight at 4°, RNA was collected by ethanol precipitation from the opaque supernatant. The RNA dissolved in 50 mM potassium acetate (pH 5.0) was made up to 1 M with regard to NaCl and allowed to stand at 4° overnight. From the mixture, aggregated RNA was centrifuged off at 15 000 $\times$ g for 1 h, and soluble RNA was collected by ethanol precipitation (1 M NaCl-soluble RNA). The aggregated RNA was washed twice with 1 M NaCl and collected (1 M NaCl-insoluble RNA). Approx. 35 mg of RNA were obtained from 100 g tissues.

Preparation of DNA. Frozen tissues were dispersed in 2 volumes of Tris–HCl (0.1 M, pH 9.0)–sodium dodecyl sulfate (1 %)–phenol (45 %)[5] with gentle stirring. The viscous suspension was centrifuged at 10 000 $\times$ g for 20 min. Treatment of the aqueous phase with Tris–sodium dodecyl sulfate–phenol followed by centrifugation was repeated 3 times. DNA precipitated by addition of 2 volumes of ethanol was transferred to 15 mM NaCl–1.5 mM tri sodium citrate, (pH 7.0). The crude DNA solution was incubated at 37° for 30 min with ribonuclease A (5 μg/ml) andribonuclease T$_1$ (1.5 units/ml) after addition of Tris–HCl (pH 7.6) to a final 10 mM. Protease (0.5 mg/ml) and sodium dodecyl sulfate (0.2 %) was added to the solution and the incubation was continued for 10 h with gentle agitation. The solution was deproteinized using the phenol treatment. Finally, jelly-like DNA precipitated by ethanol was spooled up onto a glass rod and stored in 70 % ethanol at −20° until use. Purity: DNA, 97 %; RNA, 2 %; protein, 1 %.

Separation of ribonuclease A digestion products. The reaction mixture (0.1 ml) consisting of 1 mg of RNA, 25 μg of ribonuclease A, 5 mM Tris–HCl (pH 7.6) and 1 mM EDTA was incubated at 37° for 18 h. The digest was treated with HCl at pH 1.0 overnight at 4°. Chromatography on DEAE-cellulose (Cl$^-$) column was carried out in the presence of 7 M urea[6] at neutral (5 mM Tris–HCl, pH 7.6) or acidic pH (20 mM formic acid, pH 3.8) with a linear gradient of NaCl.

Other methods. Assay of amino acid acceptance, analyses of base composition and organic phosphorus, and sedimentation analysis using a Spinco model E were carried out as described previously[4,7]. Radioactivity was counted in a Tri–Carb liquid-scintillation spectrometer as described previously[7].

RESULTS

Low-molecular-weight RNA

As shown in Fig. 1a, the Lingula 105 000 $\times$ g-supernatant RNA preparation was partially separated into two components (A and B) on DEAE-cellulose column chromatography. However, component B was scarcely obtained from the 1 M NaCl-soluble RNA preparation (Fig. 1b).

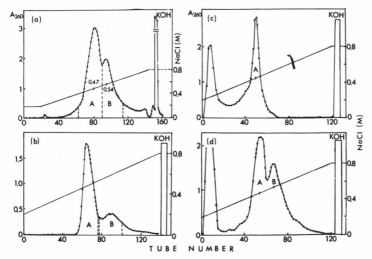

Fig. 1. Chromatography of Lingula low-molecular-weight RNA on a column of DEAE-cellulose. (a) 105 000×g supernatant RNA preparation (420 $A_{260\ nm}$ units) was applied onto a column (1.2 cm × 20 cm) of DEAE-cellulose equilibrated with Tris–Mg²⁺ buffer and eluted with a linear gradient of NaCl (0.2–0.8 M) in Tris–Mg²⁺ buffer (total volume, 500 ml). After gradient elution the remaining RNA was eluted with 0.5 M KOH. The content ratio of A, B and KOH fractions is 60, 20 and 20 %, respectively. (b) 1 M NaCl-soluble RNA preparation (3600 $A_{260\ nm}$ units) was applied onto a column (1.8 cm × 50 cm) of DEAE-cellulose. Elution was performed with a linear gradient of NaCl (0.2–0.8 M; total volume, 4.2 l) in Tris–Mg²⁺ buffer. (c, d) 100 g of Lingula tissue were ground with quartz sand at 4° and cellular components were extracted with 150 ml of solution A. The crude homogenate obtained after centrifugation at 10 000 × g for 20 min was divided into two fractions. (c) 50 ml of the crude homogenate were diluted with Tris–Mg²⁺ buffer and charged onto a column (1.0 cm × 17 cm) of DEAE-cellulose. Elution was performed with a linear gradient of NaCl (0.2–0.8 M) in Tris–Mg²⁺ buffer (total volume, 500 ml). (d) RNA was extracted from the crude homogenate by the phenol method and applied onto a column.

In order to know the cellular location of component B, the crude homogenate was chromatographed. As shown in Fig. 1c, only peak A appeared. When RNA was extracted from the homogenate, another peak B appeared with peak A (Fig. 1d). The component B was also found in the RNA preparation extracted from the 105 000 ×g pellets. From these results it is evident that component B may be associated with a nucleo-protein particle, probably ribosomes.

On rechromatography of the component A on a Sephadex G-100 column, three components (A-1, A-2 and A-3) were separated (Fig. 2a). The A-1 and A-3 components were further purified on a DEAE-cellulose column in 7 M urea (pH 3.5) (Figs. 2c and d). The elution patterns indicate that they are homogeneous in size. Component B was eluted somewhat later than Dextran blue 2000 from a Sephadex G-100 column (Fig. 2b). This elution position suggests that the molecular weight of B-RNA is close to $1 \cdot 10^5$.

Some properties of A-1, A-2, A-3 and B-RNA components

The sedimentation coefficients, $s^o_{20,w}$, at infinite dilution of 3 RNA components were determined as follows: 4.7 S for A-1 RNA, 3.7 S for A-2 RNA and 2.4 S for A-3 RNA.

21

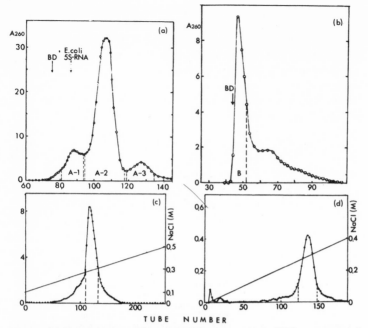

Fig. 2. (a) Fractionation of Lingula RNA component A. RNA component A (360 $A_{260\,nm}$ units) in Fig. 1b was loaded on top of the column (1.0 cm × 146 cm) of Sephadex G-100 and eluted with 0.1 M NaCl in 10 mM Tris–HCl buffer, pH 7.6. Fraction volume, 0.6 ml; flow rate, 0.2 ml/min. The content ratio is as follows. A-1 : A-2 : A-3 = 13 : 76 : 11. The symbol BD indicates a void volume of the column. The elution position of *E. coli* 5-S RNA is also indicated. (b) Purification of Lingula RNA component B. RNA component B (80 $A_{260\,nm}$ units) in Fig. 1a was fractionated by gel filtration on Sephadex G-100 column (1.0 cm × 140 cm). Other conditions as in (a). (c) Chromatography of Lingula A-1 RNA. A-1 RNA (63 $A_{260\,nm}$ units) in Fig. 2a was applied onto a column (0.3 cm × 60 cm) of DEAE-cellulose, buffered with 7 M urea–0.1 M formic acid (pH 3.5) and eluted with a linear gradient of NaCl (0.1–0.5 M) in 7 M urea–0.1 M formic acid (pH 3.5) (total volume, 200 ml). (d) Chromatography of Lingula A-3 RNA. A-3 RNA (20 $A_{260\,nm}$ units) in Fig. 2a was applied onto a column (0.25 cm × 40 cm) of DEAE-cellulose, buffered with 7 M urea–0.1 M formic acid (pH 3.5) and eluted with a linear gradient of NaCl (0–0.4 M) in 7 M urea–0.1 M formic acid (pH 3.5) (total volume, 100 ml).

TABLE I

NUCLEOTIDE COMPOSITION OF LINGULA RNA's

RNA source*	Percentage nucleotide composition**							
	Aoh	pGp	Gp	Ap	Cp	Up	Ψp	Tp
A-1	—	—	29.1	22.7	24.9	22.9	0.4	—
A-2	1.0	0.9	25.7	20.7	27.4	20.9	2.5	0.9
A-3	2.1	—	22.8	15.6	38.8	16.9	2.8	—
B	—	—	29.3	20.6	26.9	22.2	1.0	—
KOH fraction	—	—	34.5	23.8	21.7	20.0	—	—

* A-1, A-2, A-3 and B-RNA's in Fig. 2 (a–d) and KOH fraction in Fig. 1a were used for analysis.

** Average of two experiments.

22

In Table I the nucleotide compositions of RNA components are compared. The nucleotide composition of B-RNA is similar to those of A-1 and A-2 RNA's for the 4 major nucleotides, but is different from that of the KOH fraction (probably ribosomal RNA, *cf.* Table III). So, B-RNA is probably not a random degradation product of ribosomal RNA even if it is derived from them.

Amino acid-accepting ability was tested in 4 RNA components using the aminoacyl-tRNA synthetase from baker's yeast. The results in Table II show that only A-2 RNA has the ability to accept the amino acids tested.

TABLE II

<small>AMINO ACID ACCEPTANCE OF LINGULA RNA'S</small>

Amino acid acceptance assay using the aminoacyl-tRNA synthetase from baker's yeast was carried out as described previously[7]. RNA was pretreated at pH 9.5 in order to remove endogenous amino acids before assay[7]. Each figure represents counts/min of labelled amino acid bound by $A_{260\ nm}$ unit of RNA. Background (*minus* RNA) was subtracted.

RNA source*	Counts/min per $A_{260\ nm}$ unit			
	$[^{14}C]Leu$	$[^{14}C]Tyr$	$[^{14}C]Pro$	$[^{3}H]Ala$
A-1	10	37	5	28
A-2	138	1173	95	631
A-3	23	123	8	38
B	31	18	12	26
Yeast tRNA	694	6033	223	1931

* RNA preparations used are the same as described in Table I.

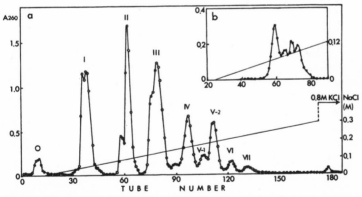

Fig. 3. (a) Chromatography of a ribonuclease A digest of Lingula A-3 RNA. Ribonuclease A digest of A-3 RNA (1 mg) in Fig. 2d was applied onto a column (0.25 cm × 30 cm) of DEAE-cellulose, buffered with 7 M urea–5 mM Tris–HCl (pH 7.5) and eluted with a linear gradient of NaCl (0–0.3 M) in the same urea buffer (total volume, 100 ml). The recovered absorbance (at 260 nm) and the percentate of each fraction in the total digest (20.7 $A_{260\ nm}$ units) are as follows: Fraction 0, 0.51 $A_{260\ nm}$ unit (2.5 %); Fraction I, 4.61 (22.2); Fraction II, 4.25 (20.5); Fraction III, 6.06 (29.2); Fraction IV, 2.42 (11.7); Fraction V-1, 0.60 (2.9); Fraction V-2, 1.61 (7.7); Fraction VI, 0.38 (1.8); Fraction VII, 0.25 (1.2). (b) Rechromatography of the pentanucleotide fraction. Fraction V-2 (1.61 $A_{260\ nm}$ units) was applied onto a column (0.2 cm × 20 cm) of DEAE-cellulose, buffered with 7 M urea–0.02 M formic acid (pH 3.8) and eluted with a linear gradient of NaCl (0–0.12 M) in the same urea buffer (total volume, 40 ml).

In order to know the homogeneity of molecular species[8], ribonuclease A digest of A-3 RNA was chromatographed on a column of DEAE-cellulose in 7 M urea at pH 7.6 (Fig. 3a). The nucleoside (Fraction o) released from the 3'-termini of A-3 RNA is adenosine. From the adenosine content (2.1% of total moles of phosphates) the chain length is calculated to be 48 nucleotides on the average. A-3 RNA contains only 1.2 % of heptanucleotide (Fraction VII), and pentanucleotide (Fraction V-2) was separated into 4 peaks by rechromatography on a AG-1 column (Fig. 3b). From these results it is concluded that A-3 RNA is constituted from heterogeneous molecular species.

High-molecular-weight nucleic acids

1 M NaCl-insoluble RNA preparation was purified by Sephadex G-200 gel filtration. The nucleotide composition of the purified high-molecular-weight (ribosomal) RNA is consistent with that of the KOH fraction (Table III).

TABLE III

NUCLEOTIDE COMPOSITION OF LINGULA DNA AND HIGH-MOLECULAR-WEIGHT RNA.

Source	$\varepsilon(P)$	Percentage nucleotide composition*				
		G	A	C	U(T)	G+C %
RNA**	8050	32.7	22.9	22.3	22.1	55.0
DNA	6080	16.2	33.6	18.8	31.4	35.0

* Average of two experiments.
** High-molecular-weight RNA purified on a Sephadex G-200 column from 1 M NaCl-insoluble RNA was used for the analysis.

Enzymatic hydrolysis of the purified DNA with pancreatic deoxyribonuclease and snake venom phosphodiesterase produced 4 usual mononucleotides (Table III). There are complementarities in A · T and G · C base pairs. The G+C content (35 %) is less than that in vertebrates but is within the range of that in invertebrates analyzed[9]. The ultraviolet absorbance of Lingula DNA per phosphorus, $\varepsilon(P)$, at 258 nm in 0.15 M NaCl–0.015 M trisodium citrate, pH 7.0, was calculated as 6080. This value is reasonable for natural double-stranded DNA[10].

DISCUSSION

Lingula A-1 RNA has a sedimentation coefficient of 4.7 S. Chemical analysis of ribonuclease A digest indicated that A-1 RNA is homogeneous in molecular species (unpublished data). A small amount of pseudouridylate (Ψp) may be derived from a contaminated tRNA. This RNA has no accepting ability for amino acids and is eluted at a position similar to *Escherichia coli* 5-S RNA on a Sephadex G-100 column. From these results A-1 RNA may be a species of 5-S RNA[11].

Lingula A-2 RNA has a sedimentation coefficient of 3.7 S. It contains adenosine, pGp and some minor bases and shows amino acid-accepting activity. Therefore, A-2 RNA can be identified as an amino acid transfer RNA. This conclusion will be confirmed in a following paper.

Lingula A-3 RNA is homogeneous in size and has a sedimentation coefficient of 2.4 S but is heterogeneous in molecular species. The 3'-end of this RNA is adenosine, the same as in tRNA, while A-3 RNA does not accept an amino acid. The nucleotide composition of A-3 RNA is distinct from other RNA components and the cytidylate content is extremely high. It remains to be resolved whether A-3 RNA is unique for Lingula or is a specific fragment produced by a nuclease from tRNA and/or other RNA's.

Another RNA component, distinct from tRNA and 5-S RNA, has been isolated from a number of animal cells[12-15]. This RNA species resembles tRNA in base composition but has a sedimentation coefficient of about 7 S and does not accept amino acids. The RNA molecule (nominally 7-S RNA) contains only a trace of Ψp and does not contain methylated bases. The mammalian 7-S RNA has been shown to be a metabolically unstable component of ribosomes[12]. Lingula B-RNA seems to belong to this category of RNA, although it is stable and its molecular size is somewhat larger. It is also likely that B-RNA may be a stable messenger RNA. In an *in vitro* protein-synthesizing system from mouse liver and Sarcoma 180 cells 7-S RNA stimulates the incorporation of amino acids into protein[12]. The question has not been answered why B-RNA is found attached to nucleo-protein particles (probably ribosomes).

ACKNOWLEDGEMENTS

The authors thank Dr. M. Kuwano, Kyushu University, and Mr. K. Kuwano, Mikuni Sangyo at Fukuoka, for very kindly arranging to supply Lingula and for their continuing interest, and Dr. S. Takemura of our Institute for his helpful discussion. This work was supported in part by a grant from the Ministry of Education.

REFERENCES

1 E. O. DODSON, *Evolution: Process and Product*, Reinhold, New York, 1962, p. 148.
2 L. H. HYMAN, *The Invertebrates*, McGraw-Hill, New York, 1940,
3 M. KUME AND K. DAN, *Embryology of Invertebrates* (in Japanese), Baihu-kan, Tokyo, 1957, p. 191.
4 N. SHIMIZU, K. MIURA AND H. AOKI, *J. Biochem.*, 68 (1970) 265.
5 H. SAITO AND K. MIURA, *Biochim. Biophys. Acta*, 72 (1963) 619.
6 R. V. TOMLINSON AND G. M. TENER, *J. Amer. Chem. Soc.*, 84 (1962) 2644.
7 N. SHIMIZU, H. HAYASHI AND K. MIURA, *J. Biochem.*, 67 (1970) 373.
8 R. W. HOLLEY, G. A. EVERETT, J. T. MADISON AND A. ZAMIR, *J. Biol. Chem.*, 240 (1965) 2122.
9 H. A. SOBER, *Handbook of Biochemistry*, The Chemical Rubber Co., 1968, Section H.
10 R. L. SINSHEIMER, *J. Mol. Biol.*, 1 (1959) 43.
11 G. ATTARDI AND F. AMALDI, *Ann. Rev. Biochem.*, 39 (1970) 183.
12 J. D. WATSON AND R. K. RALPH, *J. Mol. Biol.*, 22 (1966) 67.
13 J. PENE, E. KNIGHT, JR. AND J. E. DARNELL, JR., *J. Mol. Biol.*, 33 (1968) 609.
14 T. NAKAMURA, A. W. PRESAYKO AND H. BUSCH, *J. Biol. Chem.*, 243 (1968) 1368.
15 J. D. WATSON AND R. K. RALPH, *Biochem. Biophys. Res. Commun.*, 24 (1966) 257.

Increasing the Multiplicity of Ribosomal RNA Genes in Drosophila melanogaster

Kenneth D. Tartof

In the eucaryotes thus far studied the multiple copies of the 18S and 28S RNA genes are clustered at the nucleolus organizer (NO) locus (*1–3*). Experiments with both *Xenopus* and *Drosophila* have shown that the amount of 18S and 28S ribosomal RNA (rRNA) hybridizable per unit DNA is directly proportional to the dosage of the NO segment per genome (*2, 3*). However, in the course of investigating the redundancy of the rRNA genes in *Drosophila melanogaster*, I have discovered several cases which contradict this observation. When flies carry only one NO region (Fig. 1A), on their X chromosome, there are approximately 150 more rRNA genes per X as compared to the value when the same wild-type X chromosome is present as part of the usual two doses in X/X females. This striking increase occurs during the development of these individuals, probably involves more than one cell type, and cannot be due to gene recombination between homologous chromosomes.

It appears to involve a mechanism whereby the rRNA gene cluster of a standard wild-type X chromosome undergoes disproportionate replication. This phenomenon may be related to the magnification phenomenon described by Ritossa (4).

When X/X wild-type females [from the wild-type stock Ore-R M (5)] are mated to males that have their X and Y chromosomes attached (symbolized here as $\overline{XY}$, Fig. 1B), X/$\overline{XY}$ females and X/0 males are produced (6, 7). DNA's from both parents and progeny of this cross were isolated, and the percentage of the DNA complementary to rRNA was determined as indicated (Table 1). In experiment 1 (Table 1) the X chromosome of the wild-type female contains approximately 255 rRNA genes. When this same X was combined with an $\overline{XY}$ chromosome in X/$\overline{XY}$ females, the number of rRNA genes was the sum of the two chromosomes. However, when present as a single dose in X/0 males, this X then contained 430 rRNA genes, or an increase of approximately 175 rRNA genes per X. This increase in the number of rRNA genes per X in X/0 males is not the result of erratic variability in the DNA-RNA hybridization technique or the result of genetic contamination of the *Drosophila* stocks for the following reasons. (i) A second independent isolation and hybridization of DNA from the wild-type females used in experiment 1 (Table 1) gave a value for the percentage of DNA hybridized of 0.462 ± 0.020 as compared to 0.445 ± 0.022, confirming the reproducibility of the measurements obtained by the hybridization technique. Moreover, two separate ways of calculating the number of rRNA genes per X are in close agreement. Dividing the percentage of DNA hybridized of X/X females by 2 gives a value of 255 rRNA genes per X; whereas subtracting the percentage of DNA hybridized of $\overline{XY}$ males from the percentage of DNA hybridized of X/$\overline{XY}$ females gives 275 rRNA genes per X. The additivity of the number of rRNA

genes in the X and $\overline{XY}$ chromosomes serves as a useful independent check on the reproducibility of the extraction and hybridization of the DNA. (ii) Brain squashes from six larvae of X/0 and X/$\overline{XY}$ progeny were also made and the metaphase nuclei were examined. In all cases the male larvae contained a single X and three pairs of autosomes, and the female larvae contained one X, one $\overline{XY}$, and three pairs of autosomes. Therefore, the parents and progeny in experiment 1 (Table 1) do indeed possess the indicated genotypes free of any extra contaminating chromosomes. Also, since X/0 males are characteristically sterile because they lack a Y chromosome, several thousand X/0 males produced in experiment 1 (Table 1) were randomly selected and tested for their infertility. In all cases such X/0 males were sterile.

These experiments have also been repeated with a wild-type stock [Ore-R S stock (5)] whose X chromosome is of separate origin and an attached XY chromosome that has been carried in a stock separate from the one used in experiment 1 for several years—symbolized here as $\overline{XY'}$ (8). The results are summarized in experiment 2 (Table 1). Again, the X chromosome is additive with the $\overline{XY'}$, but an increase of approximately 130 rRNA genes per X occurs when this X is present as a single dose in X/0 males as compared to X/X females. The genotype of these X/0 males was verified by testing their sterility; cytological examination of X/0 male and X/$\overline{XY'}$ female larvae also confirmed their genotypes.

The use of yet another X chromosome derived from the X used in experiment 2 (Table 1) but maintained in X/$\overline{XY}$ females as a separate line [stock A01-1 (6, 7)] for several years also demonstrates the increase in rRNA genes per X when it is present as a single dose in X/0 males. In this experiment $\overline{XY}$ males were mated to X/$\overline{XY}$ females, and the progeny of $\overline{XY}$ males, X/$\overline{XY}$ females, and X/0 males (6, 7) were separated and their DNA was iso-

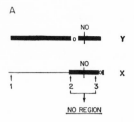

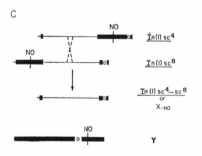

Fig. 1. Various rearrangements of the sex chromosomes of *Drosophila melanogaster* (7). (A) In addition to three sets of autosomes the male carries one X and one Y chromosome while the female contains two X's. The X and Y each contain one nucleolus organizer (NO) as indicated. The heavily darkened part of the X indicates the hetrochromatic region of this chromosome while the thin line represents the euchromatic portion; the Y is entirely heterochromatic. The open circle (○) designates the centromere. (B) A chromosome with the X and Y attached together can be synthesized whereby the short arm of the Y (Y^s) is linked to the left arm of an inverted X chromosome, and the long arm of the Y (Y^L) to the right arm of this X. This chromosome has the genetic symbol $Y^s X \cdot Y^L$, *In(1)EN* (7) but for convenience is abbreviated here as $\overline{XY}$. It contains one (or possibly two) NO clusters in the left heterochromatic arm (7). (C) An X chromosome can also be synthesized which lacks its NO and most of the surrounding heterochromatin. A female is obtained which contains one X chromosome designated *In(1)sc⁴* which contains an inversion between points 1 and 2 (see A) while the other X is inverted between points 1 and 3 and is designated as *In(1)sc⁸*. With appropriate markers on each X, such a female is mated to a male and the recombinant chromosome, *In(1)sc⁴-sc⁸* or X₋NO, is obtained in a male which contributes the NO of its Y for viability. The X₋NO chromosome lacks the NO region which is defined as that chromosome segment between points 2 and 3 of the inversions.

lated and hybridized to rRNA. The results are given in experiment 3 (Table 1). The X chromosome in X/$\overline{XY}$ females contains about 103 rRNA genes, but in X/0 males it has 248 such genes. This represents an increase of approximately 145 rRNA genes per X. The number of rRNA genes in this particular X is rather low compared to X chromosomes used in experiments 1 and 2 (Table 1). The precision of the hybridization technique (as discussed above) is such that these differences are significant and indicate, in agreement with a previous report (9), that different stocks may have widely different optima for rRNA gene redundancy. However, in this case as well, the rRNA gene content of the X is increased significantly when in the X/0 condition.

These three experiments demonstrate that a wild-type X chromosome can increase its content of rRNA genes when present as a single dose in X/0 males relative to the number of rRNA genes per X in the X/X or X/$\overline{XY}$ female parent. Since this increase in rRNA genes occurs in X/0 males, it cannot be explained by gene recombination between homologs. However, an explanation involving sister chromatid exchange remains a possibility.

It was also of interest to determine whether a wild-type X would increase its content of 18S and 28S rRNA genes when opposite an X chromosome that lacks its NO region (X_{-NO}, Fig. 1C). Wild-type X/X females were mated to X_{-NO}/Y males (10), and the progeny, X_{-NO}/X females and X/Y males, recovered. The percentage of DNA hybridizable to 18S and 28S rRNA was determined on both parents and progeny, and the results are summarized in experiment 4 (Table 1). The rRNA genes of the X chromosome from X/X females are additive with those of the Y from X_{-NO}/Y males. However, in X_{-NO}/X females the X chromosome has increased its content of rRNA genes per X from about 255 to 395, an increase of approximately 140 rRNA genes per X relative to the X/X female.

These results indicate that, when the NO region (Fig. 1A) of the wild-type X chromosome occurs as a single dose in either X/0 males or X_{-NO}/X females there is a dramatic increase of approximately 150 rRNA genes per X relative to the number of such genes per X in the X/X female parent. I propose that this increase occurs as the result of disproportionate replication of the rRNA genes at some point in the course of the developmental cycle.

In contrast to the above, males whose Y chromosome has been exposed for many years to an X_{-NO} chromosome (in X_{-NO}/Y males) have half the number of rRNA genes as X/X females or X/Y males (3; experiment 4). This could be interpreted to mean that the rRNA genes in the Y chromosome are incapable of disproportionate replica-

tion. However, the number of rRNA genes in the Y before exposure to the X_{-NO} is not known, and the possibility remains that there has been disproportionate replication to the present level as the result of exposure to the X_{-NO} chromosome.

One obvious question is whether the X_{-NO}/X females can pass their increased rRNA gene number to successive generations. There is experimental evidence (11) that when the X_{-NO}/X females of experiment 4 (Table 1) were again mated to X_{-NO}/Y males (10) the resulting progeny of X_{-NO}/X daughters gave a value of the percentage of DNA hybridized per X nearly equal to that of their mothers while X/Y sons showed a percentage of DNA hybridized per X approximately equal to that of the standard wild-type X chromosome in X/X females (Ore-R M stock; 5). Thus, the increased rRNA gene number persisted in those individuals of subsequent generations in which only a single *NO region* was present and did not persist in those individuals of subsequent generations in which the wild-type NO of a Y chromosome was present.

Flies partially deficient in rRNA genes have much smaller bristles and develop more slowly than the wild type, a phenotype called bobbed, *bb* (7). Such *bb* mutants have the characteristic of reverting after a few generations to the wild-type phenotype, with a concomitant increase in the number of rRNA genes (3, 4, 7). It was suggested that this phenomenon, termed magnification, might be due to disproportionate gene replication (4). The striking increase in rRNA genes reported here occurs during the ontogeny of a single generation and is consistent with this view. Furthermore, my results suggest that the extent of disproportionate replication of the rRNA genes in an X chromosome could be inversely proportional to the number of rRNA genes in the opposite homolog. This would allow conditions of low rRNA gene redundancy to revert rapidly and yet

Table 1. Hybridization of ^{3}H-labeled 18S and 28S RNA to DNA of various genotypes. Females from the Ore-R M stock, Ore-R S stock, or A01-1 stock were mated to males (described in the text), and their progeny was separated according to genotype. Methods of maintaining *Drosophila melanogaster* stocks, purification of ^{3}H-labeled 18S and 28S RNA (45,000 count/min per microgram) and DNA-RNA hybridization have been described (*13*), except that, in the present experiments, hybridization time was shortened to 3 hours. The reaction was complete after 2 hours and remained unchanged for at least 12 hours. Filters containing 50 to 65 μg of DNA from adult flies of the appropriate genotypes were hybridized with 8.8, 13.2, 17.6, 22.0, and 26.4 μg of 18S and 28S ^{3}H]RNA. The plateau value was calculated in the following manner. The percentage of DNA hybridized at 13.2, 17.6, 22.0, and 26.4 μg of ^{3}H]rRNA from two or three determinations was averaged, and the standard error (S.E.) was calculated from these 8 or 12 values, respectively. Flat saturation plateaus were achieved over this range [see (*13*)]. As a typical example of the number of counts per minute per filter, filters containing 65 μg of DNA from X/X female or X/0 male DNA (experiment 1) gave, on the average, approximately 13,000 and 12,000 count/min per filter, respectively. In order to compare the percentage of DNA hybridized in X/0 males or X/$\overline{XY}$ females to that in the wild type, a correction factor must be introduced to compensate for the different DNA contents of these genotypes. The X or Y chromosome accounts for 10 percent of the DNA per genome. Therefore, since the concentration of 18S and 28S RNA genes per unit DNA in X/0 males will be enriched 10 percent relative to the wild-type genotype, the percentage of DNA hybridized in such males must be multiplied by 0.9. Conversely, the percentage of DNA hybridized in X/$\overline{XY}$ females must be multiplied by 1.1 to compensate for the 10 percent dilution of 18S and 28S RNA genes per unit DNA in this genotype. The number of rRNA genes was calculated with the assumption that the molecular weight of the diploid genome of *Drosophila melanogaster* is 2.4×10^{11} and the molecular weight of 18S and 28S RNA is 2.1×10^6 (*13*). Figures in parentheses represent the number of determinations.

Genotype	DNA hybridized ± S.E.* (%)	Corrected DNA hybridized (%)	DNA hybridized per X (%)	18S + 28S RNA genes per X (No.)	Net increase of rRNA genes in X/0 or X_-NO/X genotypes (No.)
Experiment 1 (Ore-R M, X/X ♀ × $\overline{XY}$ ♂)					
P₁ { $\overline{XY}$ ♂	0.263 ± 0.022 (3)	0.263			
{ X/X ♀	0.445 ± 0.022 (2)	0.445	0.223	255	
F₁ { X/$\overline{XY}$ ♀	0.458 ± 0.010 (2)	0.504	0.241	275	
{ X/0 ♂	0.418 ± 0.020 (2)	0.376	0.376	430	175
Experiment 2 (Ore-R S, X/X ♀ × $\overline{XY'}$ ♂)					
P₁ { $\overline{XY'}$ ♂	0.211 ± 0.011 (2)	0.211			
{ X/X ♀	0.450 ± 0.016 (2)	0.450	0.225	256	
F₁ { X/$\overline{XY'}$ ♀	0.366 ± 0.018 (2)	0.403	0.192	219	
{ X/0 ♂	0.375 ± 0.030 (3)	0.338	0.336	386	130
Experiment 3 (A01-1, $\overline{XY}$/X ♀ × $\overline{XY}$ ♂)					
P₁ { $\overline{XY}$ ♂	0.263 ± 0.022 (3)	0.263			
{ X/$\overline{XY}$ ♀	0.321 ± 0.014 (2)	0.353	0.090	103	
F₁ { $\overline{XY}$ ♂	0.263 ± 0.022 (3)	0.263			
{ X/$\overline{XY}$ ♀	0.321 ± 0.014 (2)	0.353	0.090	103	
{ X/0 ♂	0.241 ± 0.019 (3)	0.217	0.217	248	145
Experiment 4 (Ore-R M, X/X ♀ × X_-NO/Y ♂)					
P₁ { X_-NO/Y ♂	0.217 ± 0.015 (2)	0.217			
{ X/X ♀	0.445 ± 0.022 (2)	0.445	0.223	255	
F₁ { X/Y ♂	0.467 ± 0.018 (2)	0.467	0.250	286	
{ X_-NO/X ♀	0.346 ± 0.018 (2)	0.346	0.346	395	140

* The value for the percentage of DNA hybridized in $\overline{XY}$ males in experiments 1 and 3 and in X/X females in experiments 1 and 4 is taken from the same set of determinations.

would also maintain the wild-type level of 18S and 28S rRNA genes at a relatively constant multiplicity.

The phenomenon of magnification is confined exclusively to males carrying mutations of the rRNA genes in both sex chromosomes (4). The significance of the system described above is that disproportionate rRNA gene replication occurs in standard wild-type X chromosomes. Moreover, the fact that this occurs in both X/0 males and X$_{-NO}$/X females suggests that more than a single cell type is involved, and thus the phenomenon may be analogous to the amplification of the rRNA genes in the oocyte of many organisms [see (12)]. Finally, it appears that in *Drosophila* disproportionate gene replication is associated with a mechanism capable of sensing a deficiency in the number of rRNA genes.

References and Notes

1. R. P. Perry, *Proc. Nat. Acad. Sci. U.S.* **48**, 2179 (1962); D. D. Brown and J. B. Gurdon, *ibid.* **51**, 139 (1964); E. H. McConkey and J. W. Hopkins, *ibid.*, p. 1197; M. L. Pardue, S. A. Gerbi, R. A. Eckhardt, J. G. Gall, *Chromosoma* **29**, 268 (1970).
2. M. L. Birnstiel, H. Wallace, J. L. Sirlin, M. Fischberg, *Nat. Cancer Inst. Monogr.* **23**, 431 (1966).
3. F. M. Ritossa, K. C. Atwood, D. L. Lindsley, S. Spiegelman, *ibid.*, p. 449.
4. F. M. Ritossa, *Proc. Nat. Acad. Sci. U.S.* **60**, 509 (1968).
5. Ore-R M and Ore-R S stocks are wild-type lines established from two different populations. These stocks as well as all others described in this report are maintained in the collection of Dr. Jack Schultz, Institute for Cancer Research, Philadelphia, Pa.
6. $\overline{XY}$ males are maintained by mating them to females that possess one $\overline{XY}$ chromosome and an X chromosome (derived from the Ore-R S stock) carrying the mutant gene, y^s (7) which causes the body color of the fly to be yellow as in X/0 males. This stock has the genetic symbol $Y^s X \cdot Y^L$, $In(1)EN/y^s$ (7). For simplicity it is referred to as the A01-1 or $\overline{XY}$/X stock. $\overline{XY}$ males mated to $\overline{XY}$/X females give rise to a progeny of $\overline{XY}$ sons that are genetically identical to their fathers and $\overline{XY}$/X daughters that are genetically identical to their mothers. X/0 males are also produced but they are sterile, lacking a Y chromosome for fertility. $\overline{XY}$/$\overline{XY}$ females usually do not survive (less than 1 in 2000) in this particular stock, and those that do are frequently sterile. Thus, the only viable and fertile flies produced in a stock of $\overline{XY}$ males and $\overline{XY}$/X females are $\overline{XY}$ males and $\overline{XY}$/X females that are genetically identical to their parents. The $\overline{XY}$ males of this stock were used in experiments 1 and 3.
7. D. L. Lindsley and E. H. Grell, *Genetic Variations of* Drosophila melanogaster (Carnegie Institution of Washington, Publ. No. 627, Washington, D.C., 1968).
8. $\overline{XY}'$ males were derived from the $\overline{XY}$ males described above (5) in 1966 and maintained as a separate line by mating them to females that have their own X chromosomes attached ($\overline{XX}$, attached X). The progeny of the mating of $\overline{XY}'$ males to $\overline{XX}$ females are $\overline{XY}'$ sons and $\overline{XX}$ daughters that have genotypes identical to their fathers and mothers, respectively. The $\overline{XY}'$/$\overline{XX}$ genotype is lethal. This stock is described by the symbol, $Y^s X \cdot Y^L$, $In(1)sc^{4L}sc^{8R}$, w^{48h} (7).
9. F. M. Ritossa and G. Scala, *Genetics* **61** (Suppl.), 305 (1969).
10. X$_{-NO}$/Y males are maintained in a stock whereby they are mated to a female that possesses a Y and an attached X which consists of two X chromosomes tandemly linked together, both of which lack the *NO region* (Fig. 1). This stock is designated as $In(1)sc^{4L}sc^{8R}$, y sc^4sc^8 cv v $B/C(1)DX$, y f/B^sY (7).
11. K. Tartof, in preparation.
12. J. G. Gall, *Genetics* **61** (Suppl.), 121 (1969).
13. K. D. Tartof and R. P. Perry, *J. Mol. Biol.* **51**, 171 (1970).
14. I thank Drs. Jack Schultz and Robert P. Perry for stimulating discussion and Dana Tartof for technical assistance. This work was supported by PHS postdoctoral fellowship grant CA-40014-01, NSF grant GB-7051 (to Dr. Robert P. Perry), NIH grants CA-01613 (to Dr. Jack Schultz), CA-06927 and RR-05539 (to the Institute for Cancer Research), and and an appropriation from the Commonwealth of Pennsylvania.

31

RNA IN CYTOPLASMIC AND NUCLEAR FRACTIONS OF CELLULAR SLIME MOLD AMEBAS

S. M. COCUCCI and M. SUSSMAN

INTRODUCTION

Recent studies of RNA metabolism in mammalian cells have been facilitated by the application of rapid and gentle methods for separation of the cell contents into nuclear and cytoplasmic components (Prescott et al., 1966; Penman, 1966; Penman et al., 1968; Soeiro et al., 1968). The present study describes a corresponding fractionation procedure suitable for cellular slime mold amebas. The level of cross-contamination between the nuclear and cytoplasmic fractions is shown to be very low. The conditions are gentle enough to preserve the polysomal associations in the latter, and these associations have been systematically examined in cells growing exponentially, cells in the stationary phase, and cells embarked on the morphogenetic program leading to the construction of fruiting bodies. Some studies of RNA distribution between the nucleus and cytoplasm of cells that had incorporated uridine-^{3}H over brief or sustained time periods are described. The results shed some light on the ontogeny and decay of ribosomes and rRNA during growth and morphogenesis.

METHODS

Organism and Growth Conditions

Dictyostelium discoideum, strain Ax-1, was grown at 22°C with shaking in 70 ml volumes of sterile liquid

32

medium referred to hereafter a HL-5 and contains he following: glucose (16 mg/ml); proteose peptone '14 mg/ml); yeast extract (7 mg/ml); $Na_2HPO_4 \cdot 7H_2O$ (0.95 mg/ml); KH_2PO_4 (0.5 mg/ml). Growth was exponential with a doubling time of about 12 ar and a stationary phase yield of $1-2 \times 10^7$ cells/ml (about 1-2 g dry weight per liter).

Strain Ax-1 was derived from the parent strain NC-4 (which can grow only in association with bacteria), by serial selection in a medium like the above but supplemented with liver concentrate and embryo extract (Sussman and Sussman, 1967). Sustained serial passage in liquid where fruiting body construction is impossible tends to introduce morphogenetically deficient mutants into the population. To avoid this, the stock was plated clonally at monthly intervals on nutrient agar with *A. aerogenes*. A clone that had constructed normal fruiting bodies was selected, and spores, free of bacteria, were used as inoculum in the sterile liquid HL-5 medium for the next cycle of serial passages.

Conditions of Morphogenesis

Cells were harvested from the growth medium by centrifugation at ca 500 g for 2-3 min, washed once in cold water, and resuspended in cold water at a density of 1×10^8 cells/ml. Aliquots of 0.5 ml were dispensed evenly on 2 in. Millipore membrane filters that rested on absorbent pads saturated with 1.4 ml of lower pad solution (LPS), a solution containing 1.5 mg/ml KCl, 0.5 mg/ml $MgCl_2$, 0.5 mg/ml streptomycin sulfate, and 0.04 M phosphate pH 6.5, inside 60 mm plastic Petri dishes. Under these conditions the cells proceed through the morphogenetic sequence synchronously and construct fruiting bodies over a 24 hr period (Sussman, 1966). Cells could be recovered quantitatively from a filter simply by immersing it in a few milliliters of cold water and placing the test tube on a Vortex mixer for a few seconds.

Separation of Nuclear and Cytoplasmic Fractions

A sample of 10^8 cells was suspended in 2 ml of cold HMK[1] solution containing 5% (w/v) sucrose and 4% (v/v) of the detergent Cemulsol NPT 12 (Societé des Produits Chimiques de Synthèse, Bezons, Seine-et-Oise, France). The suspension was agitated for 1 min on a Vortex mixer, incubated for 9 min in an ice bath, and then diluted with 2 ml of cold HMK containing 22% sucrose and 4% NPT 12. The suspension was examined briefly at this time under phase optics at 1250 $\times$ magnification to check the efficiency of cell

---

[1] HMK solution contains: 0.05 M HEPES buffer (N-2-hydroxyethylpiperazine-N'-2-ethane sulfonic acid, Calbiochem) adjusted to pH 7.5 with NH_4OH; 0.04 M $MgCl_2$; 0.02 M KCl.

breakage. The incidence of whole cells was routinely less than 4×10^{-4}. The nuclei were collected by centrifuging at 1000 g for 5-10 min and washed once in a 1:1 mixture of the 5 and 22% sucrose HMK-solutions. The supernatant from the first spin was further fractionated by centrifuging for 5-10 min at 10,000 g to yield a pellet called the *particulate cytoplasmic fraction* and a supernatant called the *soluble cytoplasmic fraction*.

A rapid fractionation procedure, a variant of the above method, was used in order to avoid significant breakdown of polysomes in the soluble cytoplasmic fraction. The cells were suspended in the first detergent solution, agitated on the Vortex mixer for 1 min, immediately diluted 1:1 with the second detergent solution, centrifuged at 10,000 g for 1.5 min, and the supernatant was immediately diluted 1:1 with HMK solution and centrifuged in a sucrose density gradient as described in the legend to Fig. 3.

RESULTS

Distribution of Radioactivity after Incorporation of Uridine-^3H

Cells growing exponentially in liquid medium were exposed to uridine-^3H for a period of 24 hr (2 generations). Sister cells were exposed to uridine-^3H for 15 min. The cells were harvested, separated into nuclear and cytoplasmic fractions, treated with sodium dodecyl sulfate (SDS), and centrifuged in 13-23% linear sucrose gradients. Fig. 1 shows the profiles of optical density and trichloroacetic acid (TCA)-insoluble radioactivity for samples of 10^8 cells.

Fig. 1 (left) summarizes the data for cells labeled with uridine-^3H during a 24 hr period. The radioactivity profiles show great disparities in several regions of the gradients and indicate that cross-contaminations must have been relatively slight. The OD_{260} traces of both the particulate and soluble cytoplasmic fractions show the characteristic rRNA peaks with sedimentation coefficients variously calculated as 16 and 23S (Ashworth, 1966) or 17 and 25S (Ceccarini et al., 1968). The radioactivity profiles appear to coincide with the OD_{260} profiles, and the levels of radioactivity associated with the peaks are in the ratio that would be expected for the two kinds of rRNA molecules. However, the profiles observed in the nuclear fraction appear to differ from the above profiles in two respects. First, the OD_{260} profiles do not correspond precisely. The main peak in the nuclear fraction is slightly heavier than the 23-25S rRNA of the cytoplasm. This difference is considered to be significant.

33

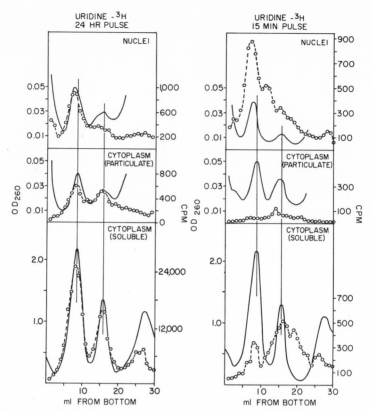

FIGURE 1 (*Left*) To a log phase culture (15 ml) of *D. discoideum* at a density of about 2×10^6 cells/ml was added 50 μCi of uniformly labeled uridine-^{3}H (27 Ci/mmole). Incubation was continued for 24 hr by which time the density of each culture was 1×10^7 cells/ml. The culture was harvested by centrifugation at 2000 g for 1.5 min. A sample containing 10^8 cells was fractionated as described in Methods. The nuclear and particulate cytoplasmic fractions were washed once, and each fraction was suspended in 0.9 ml of H_2O plus 0.1 ml of 10% SDS. The soluble cytoplasmic fractions were precipitated in 90% cold ethanol (to reduce the level of the alcohol-soluble detergent that has a high UV absorbance), and the pellet was suspended in 2.7 ml of H_2O plus 0.3 ml of 10% SDS. Samples of 1 ml were layered over linear gradients prepared from solutions containing 13–23% sucrose (w/w), 0.1 M NaCl, 0.01 M Tris buffer pH 7.4, 0.5% SDS and were centrifuged for 17 hr at 22500 rpm in an SW-25 rotor. The tubes were pierced and emptied from below, with a finger pump, through the flow cell of a Gilford recording spectrophotometer (Gilford Instrument Laboratories Inc., Oberlin, Ohio) 1-ml fractions were precipitated with 10% TCA, collected on membrane filters, washed, and counted in liquifluor scintillation fluid. (*Right*) A 15 ml culture of log phase cells (5×10^6 cells/ml) was exposed to uridine-^{3}H (12 Ci/mmole, 33 μCi/ml) for 15 min, harvested, and treated as described above. The solid lines represent OD_{260} traces (meaningless at the tops of the gradients because of the high absorbance of the detergent at 260 mμ). The dotted lines and circles represent TCA-insoluble radioactivity. All data including those for the cytoplasmic fractions are given as amounts per 10^8 cells.

Each set of three fractions was centrifuged in the same rotor and thus mutually serve as external references The difference was consistent in the experiments shown in Fig. 1 (and in repeat experiments not shown here), while in the replicate centrifugations of any one such fraction the positions of each of the peaks with respect to the meniscus agreed very closely. Second, there is a great disparity between the absolute levels of radioactivity associated with the rRNA peaks in the soluble cytoplasmic fraction on the one hand and the two peaks observed in the nuclear frac-

34

ion on the other, as well as disparate relative levels of radioactivity between the nuclear fraction peaks themselves.

The data suggest that, as in HeLa cells (Penman, 1966), the isolated *D. discoideum* nuclei are deficient in intact ribosomes and consequently in the capacity to synthesize proteins. The data shown in Table I support this conclusion. In cells exposed briefly to a mixture of ^{14}C-labeled amino acids, the TCA-insoluble radioactivity associated with the nuclei was less than 3% of the radioactivity found in the cytoplasm.

The distribution of radioactivity after a 15-min pulse of uridine-^{3}H is shown in Fig. 1 (right). About 60% of the total radioactivity is associated with the nucleus, a large proportion coincident

TABLE I

Amino acid incorporation in the nuclear and cytoplasmic fractions. 5 ml of a culture at a density of 1.5×10^7 cells/ml was incubated for 10 min with 20 μCi of a ^{14}C-amino acid mixture (New England Nuclear Corp., Boston, Mass.). The cells were harvested and fractionated, and the fractions were assayed for TCA-insoluble radioactivity.

Fraction	Total TCA-insoluble radioactivity
	cpm
Nuclei	160
Cytoplasm (particulate)	295
Cytoplasm (soluble)	6160

with the 26–27S OD$_{260}$ peak or perhaps even slightly heavier, a lesser proportion in the 16–24S region, a shoulder at 16S, a small 4S peak, and a small amount of very heavy material. Most of the remaining 40% is associated with the soluble cytoplasmic fraction, and the distribution of this radioactivity has only limited correspondence with the OD$_{260}$ trace. There are a small 23–25S peak, a pronounced 16–17S peak, a considerable proportion in the 4–16S region, and a small 4S peak. Most of this material must be unstable in view of the different distribution found after sustained exposure to uridine-^{3}H (Fig. 1, left). This result is in agreement with the results of corresponding experiments previously carried out with *Polysphondelium pallidum* (R. Sussman, 1967) in which total cell RNA was examined.

The Level of Polysomes during Growth and Morphogenesis

Soluble cytoplasmic fractions prepared from cells growing exponentially in liquid medium were centrifuged in sucrose gradients for examination of their polysomal profiles These profiles are shown in Fig. 2 *A* and *B*. A sharp peak of monosomes comprising about 20%[2] of the total ribosomal population was observed near the meniscus, and the remaining 80% was distributed through the polysomal region. Treatment with pancreatic RNase in the cold converted about 95% of the polysomes into monosomes. Addition of 0.5% sodium deoxycholate to the extracting medium did not alter the profile, thus eliminating the possibility that the heavy fraction was associated with membranous fragments. The profile is similar to that observed in log phase HeLa cells (Penman et al., 1963; Latham et al., 1965).

When the cells left the log phase and began the last generation of growth at a progressively decreasing rate (Fig. 2 *C*), the proportion of polysomes fell rapidly, and the proportion of monosomes rose. The total ribosomal population increased by about 20% as the cells grew slightly larger. When the cells entered the stationary phase at a density of about 2×10^7/ml, the proportion of polysomes had fallen to 40–45% (Fig. 2 *D*). Sister cells were harvested at this time and dispensed on membrane filters (see Methods) so that they could begin fruiting body construction. By 3 hr the level of polysomes had risen dramatically to approximately 65% of the total (Fig. 2 *E*), and by 6 hr the proportion was back to 80%. Meanwhile the total ribosomal population per cell had fallen back to the approximate level characteristic of log phase amebas. If the cells were not harvested but instead were left in the nutrient broth to complete the stationary phase, the polysome level fell to about 15% after which the cells began to die in significant numbers.

Fig. 3 summarizes an experiment in which the cells were harvested directly from the log phase and dispensed on membrane filters in order to begin morphogenesis. The level of polysomes did not fall but was maintained at about 80% throughout, even as late as 20 hr by which time the fruiting bodies were almost complete. Thus the fall in

[2] Relative proportions of polysomes and monosomes were determined by cutting graphs into appropriate segments and weighing them.

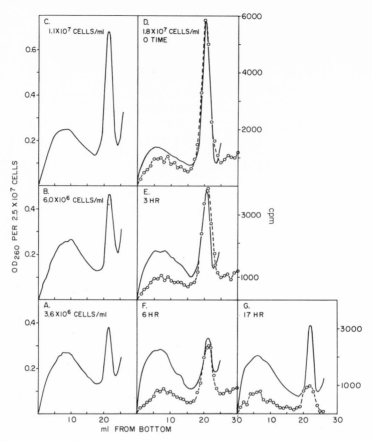

FIGURE 2   Polysomal profiles of cells in various stages of growth and morphogenesis (*A–C*). Cells were harvested from cultures that had attained the cell densities noted. Soluble cytoplasmic fractions were prepared by the rapid procedure described in the Methods section. Aliquots of 2 ml (each containing material from 2.5×10^7 cells) were layered over 28 ml of a linear 7–42% sucrose gradient (w/v) in HMK solution and centrifuged for 2 hr at 23,500 rpm in an SW-25 rotor at 3°C. The tubes were pierced and emptied from below, with a finger pump, through the flow cell of a Gilford recording spectrophotometer. (*D*) Prior to harvesting, the cells had been grown for 48 hr (4 generations) in the presence of uniformly labeled uridine-³H, specific activity = 12 Ci/mmole, 1.5 μCi/ml. The cells were washed once by centrifuging for 1.5 min at 1500 g before preparing the soluble cytoplasmic fraction. After sucrose density centrifugation as described above, the drops issuing from the spectrophotometer flow cell were collected in 1 ml fractions and immediately precipitated with 10% TCA acid in the presence of carrier bovine serum albumin. The precipitates were collected on membrane filters, washed with TCA, dried, and counted in liquifluor scintillation fluid. (*E–F*) The cells (sisters of those employed in *D*) were washed by centrifugation, resuspended in LPS solution containing 10^{-3} M cold uridine, and aliquots of 2.5×10^7 cells were dispensed on filter papers over support pads saturated with the LPS-uridine solution (see Methods section). At 3 hr and again at 6 hr, the cells from four filters were harvested and treated as in *D*. (*G*) In a separate experiment cells were harvested from filters after 17 hr of development and treated as described above. The solid lines are the OD₂₆₀ traces. The open circles and dotted lines represent cpm.

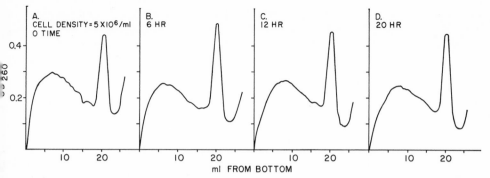

FIGURE 3   Cells were harvested from the culture medium when they had reached a density of 5×10^6 cells/ml. The soluble cytoplasmic fraction that was prepared from one sample was layered over a linear 7–42% sucrose (w/v) gradient in HMK solution and centrifuged for 2 hr at 23,500 rpm in an SW-25 rotor at 3°C; then it was collected and monitored as described in the legend to Fig. 2. The remaining cells were dispersed on filters (as described in the Methods section) in order to begin morphogenesis. Samples were harvested at the times noted and treated as above.

olysomal levels observed in Fig. 2 is not obligaory but is a specific consequence of the cells' ntrance into the stationary phase under conitions that make the construction of fruiting ·odies impossible.

The Turnover of Ribosomes during Morphogenesis

R. Sussman, studying axenically grown *Polyphondelium pallidum*, has shown that rRNA synhesized during vegetative growth is degraded luring morphogenesis and replaced by newly ynthesized rRNA having the same base composiion and the identical capacity to hybridize with *. pallidum* DNA (R. Sussman, 1967). The turnver rate was such that, over the entire morphoenetic sequence, roughly two-thirds of the old RNA disappeared and was replaced by a quanity of new rRNA roughly equal in amount to the emaining old rRNA.

This analysis was extended by examining the RNA content of *D. discoideum* ribosomes during norphogenesis. The cells harvested upon entrance nto the stationary phase (Fig. 2 D) had been rown for 48 hr (4 generations) in the presence f uridine-³H, and the radioactivity profile is een to coincide with the OD trace. Sister cells ad been washed and dispensed on membrane lters resting on support pads saturated with ·PS containing an excess of cold uridine. By 3 r (Fig. 2 E) the absolute level of radioactivity in ne polysome region had fallen as old ribosomes

disappeared, and the specific radioactivity had decreased dramatically because new ribosomes containing unlabeled rRNA had entered the polysomal region. The absolute level of radioactivity in the monosome peak had decreased much more markedly than in the polysome region as old ribosomes were degraded, but the specific radioactivity did not decrease, i.e. no new ribosomes containing unlabeled rRNA entered this pool. By 6 hr (Fig. 2 F), both total and specific radioactivity had further decreased in the polysomal region due to the disappearance of old polysomes and the appearance of new ones. The total radioactivity of the monosomes had again decreased more markedly than that of the polysomes, but the specific radioactivity had fallen only slightly (although probably significantly). In a separate experiment, samples taken at 13.5 and 17 hr displayed progressive decreases to as low as 30% of the initial specific radioactivities of both polysomes and monosomes (Fig. 2 G). It should be noted, however, that in the later samplings the decreases in specific radioactivity of the monosomes varied considerably possibly due to increased fragility of a polysomal contingent.

In Fig. 4 the change in the specific radioactivity of the combined ribosomal population has been plotted as a function of time. It is roughly first order with a half-time constant of 12 hr. Thus by the end of fruiting body construction (24-hr), it would appear that at least 75% of the old ribosomes have disappeared and have been

37

replaced by new ones whose RNA was synthesized during the morphogenetic sequence. This value is minimal since it assumes that no recycling whatsoever occurred.

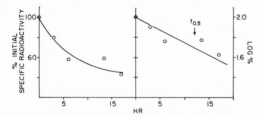

FIGURE 4 The decay of ribosomal RNA during morphogenesis. The ratios of total radioactivity: total OD_{260} were calculated from the data given in Fig. 2 D–F and also from data obtained from a second experiment (not shown) in which prelabeled cells were sampled at 0 time, 13.5, and 17 hr. Left curve shows a linear plot of the relative specific radioactivities (as percentages of the initial values) as a function of the time at which the cells were harvested. Right curve is a semilog plot of the same data.

Persistence of RNA (and Ribosomes) durin Exponential Growth

Cells were allowed to grow for 2 generations i medium containing uridine-^{3}H in order that th RNA be uniformly labeled (see Figs. 1 and 2) They were then harvested, washed free of exoge nous label, and resuspended at a density of 2 × 10^6 cells/ml in HL-5 medium containing exces $(10^{-3}$ M) cold uridine, a concentration that de creases uptake of exogenous label to an insignifi cant level under these conditions. As seen in Fig 5 (left), they continued to increase exponentiall for more than 2 generations and then grew at diminished rate as they approached the stationar phase. At intervals, samples of 2 × 10^6 cells wer taken to determine TCA-insoluble radioactivity Fig. 5 shows the cpm/ml of culture as per cent o the initial value (the latter being the value at th time the washed cells were suspended in col medium). The cells appeared to have retained a least 90% of the label even after 2 generations and the preincorporated label did not disappea

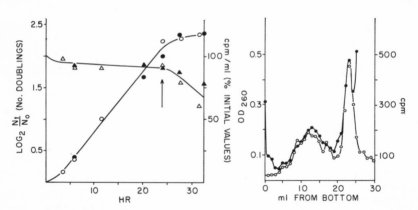

FIGURE 5 Persistence of RNA during exponential growth. (*Left*) Log phase amebas at an initial density of 5 × 10^5 cells/ml were the inoculum for 20 ml aliquots of HL-5 medium containing 50 µCi uridine-^{3}H (27 Ci/mmole) and were incubated for 24 hr until they reached a density of 2 × 10^6 cells/ml. The cells were harvested and washed in unlabeled HL-5 medium under conditions of sterility, resuspended at a density of 2 × 10^6 cells/ml in HL-5 medium containing 10^{-3} M cold uridine and 500 µg/ml streptomycin, and incubated further. At intervals thereafter, cell counts were made. Samples containing 2 × 10^6 cells were centrifuged and the pellets were examined for TCA-insoluble radioactivity. The data from two separate experiments are shown. Growth (*circles*) is plotted as Log$_2$Nt/No; TCA-insoluble radioactivity (*triangles*) is plotted in terms of cpm/ml of culture as a percentage of the initial value (the radioactivity was measured at the time the cells were removed from uridine-^{3}H medium and introduced into cold uridine medium). (*Right*) At 24 hr after being introduced into cold uridine medium (see arrow in the graph on the left), cells were harvested and samples were treated according to the procedure described in the legend to Fig. 2 to yield polysome profiles. (Closed circles, OD_{260}; Open circles, TCA-insoluble radioactivity.)

t an appreciable rate until the end of exponential growth. (In another experiment not shown here, the decrease in radioactivity was followed until it had dropped to about 50% of the initial level.) While the possibility exists that the preformed RNA was in fact destroyed and the labeled nucleotides were then recycled, one would not expect virtually total conservation as observed in Fig. 5, particularly since the presence of 10^{-3} M cold uridine in the medium has been shown to saturate the internal nucleotide pool and would therefore be expected to dilute out the labeled nucleotides.

Even if the preformed RNA were not destroyed by exponentially growing cells, the possibility remained that it had been discarded by them in terms of function. The data in Fig. 5 (right) argue compellingly against this. The distribution, in a polysomal profile, of RNA synthesized 2 generations before, representing 25% of the total RNA, was not detectably different from that of the remaining 75% of the RNA that had been synthesized subsequently.

DISCUSSION

The levels of polyribosomes have been examined in D. discoideum cells that (a) were growing exponentially, (b) had approached or entered the stationary phase, and (c) were embarked on the program of fruit construction. In exponentially growing cells, the level is very high and involves almost all of the available ribosomes. The level falls precipitously in the stationary phase but, if such cells are permitted to initiate the program of fruit construction, it rapidly rises to the original value. If this is taken as a direct reflection of the rate of protein synthesis, then D. discoideum amebas that are engaged in fruit construction may be synthesizing protein just as actively as if they were in the log phase. However, since the cells do not increase in mass during morphogenesis but in fact decrease somewhat, this high rate of protein synthesis must be more than matched by protein turnover. Wright et al. (1960) have studied protein turnover during fruit construction by examining the dilution of cell protein previously labeled with methionine-^{35}S, and they have estimated the rate as about 7% per hr. Over the 24 hr period required, this would lead to an almost quantitative replacement of preexisting protein. At least some of the replacement appears to be concerned with the preferential synthesis of specific enzymes previously absent or in low concentrations (Sussman and Sussman, 1969). The results taken together would seem to invalidate the supposition that the morphogenetic sequence is a metabolically trivial event.

The data in Figs. 2 and 4 confirm a previous finding in the sister genus Polysphondelium (R. Sussman, 1967). They show that in D. discoideum at least 75% of the RNA synthesized during the cell growth and division cycle is degraded during morphogenesis and is replaced by newly synthesized RNA. Since it is likely that the labeled nucleotides are recycled to some extent at least, this estimate of turnover is minimal. The pattern of ribosome disappearance observed in Fig. 2 indicates that newly made ribosomes go directly into the polysomal complexes and that, for at least the first 6 hr, it is the detached ribosomes (all of them old) whose RNA is preferentially destroyed. During this time the old ribosomes still in polysomal complexes disappear more slowly, presumably first entering the monosomal pool. In contrast, the label (uridine-^{3}H) in preformed rRNA persists in exponentially growing cells through at least 2 generations (24 hr) after removal of the labeled precursor from the medium and addition of excess cold uridine.

This poses an intriguing question. D. discoideum amoebae do not begin fruiting body construction until they have stopped growing and can complete the normal morphogenetic sequence in the absence of all exogenous nutrients. Clearly the cells would have a selective advantage if they were able to utilize old ribosomes and to avoid as much as possible the fabrication of new ones. The question is why do they not utilize old ribosomes exclusively.

It might be that, during morphogenesis, only newly made ribosomes can convey nascent mRNA to the cytoplasm and/or effect its translation. Studies of HeLa cells have indicated that mRNA transport into the cytoplasm does continue for some time after the cessation of ribosome synthesis (Darnell, 1968), but it should be noted that these cells were exponentially growing ones. In the present instance, it is necessary that transport be sustained for a period of 24 hr by stationary phase cells engaged in a morphogenetic program.

A second possibility is that the ribosomes synthesized during fruiting body construction constitute a new class of ribosomes that are functionally different from ribosomes present during the cell growth and division cycle with respect to their capacity to transport and/or to translate mRNA that is transcribed during fruiting body construction. In this connection it is of interest to note that, in heterokaryons of HeLa cells and hen erythrocytes (whose cytoplasm had been allowed to leak out prior to cell fusion), the previously quiescent hen nuclei synthesized RNA for the first few days but no RNA was found in the cytoplasm, whereas RNA newly made in the HeLa nuclei was found there; only after nucleoli developed in the hen nuclei and ribosome synthesis began did hen RNA appear in the cytoplasm and could hen-specific proteins be detected shortly thereafter (Harris et al., 1969).

It is also noteworthy that sea urchin and amphibian blastulae utilize preformed ribosomes exclusively and only resume ribosome synthesis after they enter the gastrula stage. While this correlation may spring merely from a quantitative requirement for additional protein synthe-sizing capacity, it could conceivably reflect th appearance of a new class of ribosomes th match the mRNA needed for the new develoj mental program (Gross et al., 1964; Brown ai Littna, 1966; Slater and Spiegelman, 196 Crippa et al., 1967).

This study was supported by grants from the Nation Science Foundation (GB-5976X), the America Cancer Society (Massachusetts Division), Inc. and a Institutional Grant to Brandeis University from NI (7044-03, Allocation No. 5).

REFERENCES

ASHWORTH, J. M. 1966. *Biochim. Biophys. Acta.* **129**: 211.

ATTARDI, G., H. PARNAS, M. L. HWANG, and B. ATTARDI. 1966. *J. Mol. Biol.* **20**:145.

BROWN, D. D., and E. LITTNA. 1966. *J. Mol. Biol.* **20**:81.

CECCARINI, C., and R. MAGGIO. 1968. *Biochim. Biophys. Acta.* **166**:134.

CRIPPA, M., E. H. DAVIDSON, and A. E. MIRSKY. 1967. *Proc. Nat. Acad. Sci. U.S.A.* **57**:885.

DARNELL, J. E. 1968. *Bacteriol. Rev.* **32**:262.

GROSS, P. R., and G. H. COUSINEAU. 1964. *Exp. Cell Res.* **33**:368.

HARRIS, H., E. SIDEBOTTOM, D. M. GRACE, and M. E. BRUMWELL. 1969. *J. Cell Sci.* **4**:499.

LATHAM, H., and J. E. DARNELL. 1965. *J. Mol. Biol.* **14**:1.

PENMAN, S. 1966. *J. Mol. Biol.* **17**:117.

PENMAN, S., K. SCHERRER, Y. BECKER, and J. E. DARNELL. 1963. *Proc. Nat. Acad. Sci. U.S.A.* **49**:654.

PENMAN, S., C. VESCO, and M. PENMAN. 1968. *J. Mc Biol.* **34**:49.

PRESCOTT, D. M., M. V. N. RAO, D. P. EVANSOI G. E. STONE, and J. D. THRASHER. 1966. Metho⁴ in Cell Physiol. D. M. Prescott, editor. Academ Press Inc., New York. **2**:131.

SOEIRO, R., M. H. VAUGHAN, J. R. WARNER, an J. E. DARNELL. 1968. *J. Cell Biol.* **39**:112.

SLATER, D. W., and S. SPIEGELMAN. 1966. *Proc. Na Acad. Sci. U.S.A.* **56**:165.

SUSSMAN, M. 1966 a. Methods in Cell Physiolog D. M. Prescott, editor. Academic Press Inc., Ne' York. **2**:397.

SUSSMAN, M., and R. R. SUSSMAN. 1969. *Symp. So Gen. Microbiol.* **19**:403.

SUSSMAN, R. R. 1967. *Biochim. Biophys. Acta.* **149**:40

SUSSMAN, R. R., and M. SUSSMAN. 1967. *Biochem Biophys. Res. Commun.* **29**:53.

WRIGHT, B. E., and M. L. ANDERSON. 1960. *Biochim Biophys. Acta.* **43**:62.

Regulation of Ribosomal RNA Synthesis and Its Bearing on the Bobbed Phenotype in *Drosophila melanogaster*

Jag Mohan and F. M. Ritossa

INTRODUCTION

The number of genes for ribosomal RNA (rRNA) in *Drosophila* is not constant. Certain individual flies have been found having as few as 60 such genes or as many as 600 (Ritossa and Scala, 1969). Phenotypically wild individuals have at least 130 genes for rRNA. The bobbed syndrome, which is pleiotropic and is generally characterized by delayed development, small bristles, and abdominal etching, is correlated with a reduction of the amount of DNA complementary to ribosomal RNA (rDNA). The number of genes for rRNA in flies showing the bobbed syndrome is lower than 130 (Ritossa *et al.*, 1966a). The results presented here elucidate some of the biochemical events underlying the bobbed phenotype.

In addition, we examined a series of *Drosophila* stocks in which the number of genes for rRNA ranges from 80 to 600 to decide whether the rRNA content per individual is or is not a function of the number of such genes. We found that it is not and can hence invoke the existence of regulatory mechanisms for rRNA synthesis.

MATERIALS AND METHODS

Growth medium. The various stocks were raised on a medium made by adding the following in grams to 1 liter of H_2O: cornmeal, 100; sucrose, 100; fresh baker's yeast, 100 (heat-killed during preparation); agar, 10; methyl p-hydroxybenzoate, 2.7.

Drosophila stocks. Wild-type Canton S was obtained from the University of Pavia. g^2 ty bb/$C(1)DX$, y f was from Pasadena. y w bb^{ds}; w^a bb^l/B^{S}Y and $C(1)DX$, y f; In (1) sc$^{4L. 8R}$, y sc^{4+8}cv v f and $C(1)DX$, y f were from the Oak Ridge National Laboratory. Ybb /y eq; Ybb su-var 5/In(1)w^{m4}; In(1)sc^8, bb^{1}cv/B^{s}Y; and C(1)DX, y f were from the Bowling Green collection. The stock car bb was obtained from a stock originally from the Oak Ridge collection. XY^L · Y^S bb was selected here from XY^L · Y^S (108–9), y^{2}su(w^a)w^a Y^L · Y^S/y v f from the collection of the University of Rome. bb(UC03) was selected from the wild-type Urbana strain.

The number in parentheses after the genetic composition, as indicated in the text and in some tables, represents the most probable percentage of rDNA of the indicated genotype.

Labeling of RNA for hybridization studies. Labeled ribosomal RNA was obtained exclusively from wild-type larvae as previously described (Ritossa *et al.*, 1966a).

RNA extraction. After repeated washing with water, the larvae were suspended in 10 volumes (w/v) of an extraction fluid which contained NaCl (0.1 *M*) and 1% sodium dodecyl sulfate (SDS). The suspension was homogenized in a glass homogenizer kept at 0°C, then 1 volume of phenol saturated with 0.1 *M* NaCl was added and the mixture was shaken for 20 minutes at 0–4°C. The aqueous layer was removed after centrifugation. The phenol treatment was repeated three more times, the first time for 20 minutes and the others for 10 minutes each. The phenol was then removed with two ether extractions, and the ether was eliminated by bubbling air through the solution. The RNA was then precipitated overnight after the addition of 2 volumes of a cold 80% ethanol containing Na acetate (0.02 *M*) in the cold.

Purification of RNA. The precipitate was taken up in a buffer (0.1 *M* NaCl; 0.05 *M* phosphate, pH 6.8) and passed through a methylated albumin column as detailed by Yankofsky and Spiegelman (1963). The ribosomal fractions were pooled and dialyzed 24 hours at 4°C against 100 vol of 2 × SSC (SSC is 0.15 *M* NaCl, 0.015 *M* Na citrate) and was stored at −30°C for further use.

Extraction and purification of RNA from ovaries. The procedure as described for the larvae was followed except that the homogenization was carried out in the presence of phenol. The fractions from MAK columns were TCA precipitated, collected on nitrocellulose filters, and counted in a liquid scintillation counter.

Determination of total nucleic acid content in various tissues. A modified procedure of the Schmidt-Thannhauser (1945) method was used as described by Munro and Fleck (1966). The various tissues, whole flies, pupae, and eggs, were homogenized in ice-cold 0.6 *N*

42

PCA and left in the ice for 20 minutes. After centrifugation the supernatant was discarded and the precipitate was washed twice with 0.2 N PCA.

The precipitate was suspended in 0.35 N KOH and incubation was carried out for 90 minutes at 37°C. The undissolved material was centrifuged, washed twice with 0.35 N KOH, and discarded. The supernatant was cooled on ice and protein and DNA were precipitated by adding cold PCA to a final concentration of 0.5 N. This precipitate was washed twice, and the washings were added to the RNA (supernatant) fraction. Water was added to make the PCA concentration 0.1 N; UV adsorption was measured at 260 mμ. A small volume of the RNA fraction was subjected to orcinol reaction.

The precipitate obtained by acidifying the alkaline digest of the tissue was suspended in 1 N PCA and treated for diphenylamine reaction as modified by Burton (1956) to estimate the DNA content.

Preparation of DNA from ovaries. The ovaries (300–400) were suspended in a solution containing 0.15 M NaCl, 2% SDS, and 0.1 M EDTA, and homogenized in a glass homogenizer. An equal volume of phenol saturated with 0.1 M NaCl was added, and the extraction was carried out. The aqueous layer was conserved. The interphase was extracted twice with phenol, and each supernatant was added to the first one. The supernatant was extracted twice with phenol followed by two more extractions with chloroform–isoamyl alcohol. The nucleic acids were precipitated overnight by the addition of two volumes of cold 80% ethanol containing Na acetate (0.02 M) in the cold.

The precipitate was dissolved in 0.35 N NaOH and RNA hydrolysis was carried out for 12 hours at 37°C. The pH of the digest was adjusted to neutral and dialysis was carried out extensively against 6 × SSC. When the DNA was extracted from adult flies, the technique was followed as previously described (Ritossa *et al.*, 1966a).

Hybridization and detection of rRNA/DNA hybrids. The method of Gillespie and Spiegelman (1965) was used.

Pulse labeling of RNA in the ovaries. Seven-day-old flies, mated in all cases with X/O (sterile) males, were used. The ovaries were dissected out in a drop of Vaseline oil under the binocular microscope and immediately transferred to Ringer solution. When 25 ovaries had been collected, they were washed 3 times in Ringer solution and incubated at 24°C in 0.15 ml of 0.1 mCi uridine-^{3}H/ml (24.4 Ci/mmole) made up in Ringer solution. During incubation, the ovaries remained covered with Vaseline oil. After incubation the ovaries were relatively freed of extracellular uridine and used for RNA extraction immediately.

In vivo labeling of RNA. Flies, 7 days old or younger, and mated in all cases, were given food of the following composition: 0.1 g agar, 2.5 g cornmeal, 2.5 g sucrose, 2.0 g fresh yeast, 20 ml water, and 2 mCi ^{32}P. The flies were allowed to eat this food undisturbed for the required periods, after which they were etherized and the ovaries were dissected out as detailed earlier. In some cases, the same ovaries were also used for pulse-labeling experiments. The "carcasses" were also used once for RNA extraction.

Determination of egg-laying rate. Fifteen to twenty-five females with, usually, the same or a greater number of wild-type males were introduced in bottles containing slides with food of the following composition: 100 ml water, 1.5 g agar, 0.7 g activated carbon, 25 g yeast, 10 g sugar, and 0.3 g methyl p-hydroxybenzoate. The slides were removed with minimum disturbance regularly after periods varying between 12 and 15 hours. The eggs were counted under a binocular microscope. The strips of food put on slides were quite large and thick (4 × 2.4 × 0.5 cm) since the flies have a tendency to lay eggs on the edges and sides.

Egg collection. A two-layer food system was used for this purpose. The bottom layer was composed of 2% agar for support, and the upper one of yeast with a small quantity of acetic acid. The eggs laid were gently removed with a camel's hair brush and washed in Ringer solution to get rid of the yeast. The dechorionization of the eggs was performed by treating them with 1% Na hypochlorite solution for 3 minutes. The eggs were washed again with Ringer solution and spread on a petri dish. A photograph was then taken which was later used for counting the number of eggs. The eggs were freed of Ringer solution and stored under ether–alcohol (3:1) at −20°C until biochemical analysis.

RESULTS

In trying to correlate the bobbed phenotype with rDNA content, the best parameter to measure is the total amount of ribosomes in flies of genotypes varying in rDNA content. Ribosome content, however, can be affected by factors like body size and, therefore, we refer to the results in terms of RNA/DNA ratios. In this case, total RNA is measured. This seems more valid because the major fraction of the RNA present in the cell is ribosomal, and furthermore, no apparent differences were seen to exist in the ratios of different RNA species between various genotypes. For DNA, the main known quantitative differences in the genotypes studied were confined to the rDNA, and they represented an extremely small fraction of the total DNA.

The RNA/DNA ratios were analyzed in genotypes carrying rDNA from nearly 0.08 to 0.6% of its total DNA; the results are listed in Tables 1 and 2. No substantial differences are seen in the ratios whether the amount of rDNA present in the flies is above or below the one found in wild-type. These results could be the consequence of the fact that: (a) there might be no differences in the amount of rRNA synthesis occurring in the various genotypes, (b) the delayed development allows the bobbed genotypes to compensate for eventual defects of rRNA metabolism, and finally, (c) the results might reflect the fact that the bobbed phenotype at the adult level is exhibited only by certain body traits, and the cells responsible for these do not constitute the majority of the body cells. In this case, even on the basis of the hypothesis that the bb phenotype is due to

TABLE 1
RNA/DNA RATIOS IN ADULT FEMALES OF SEVERAL GENOTYPES

Genotype	rRNA/DNA $\times$ 100 at saturation level	RNA/DNA
bb$^+$/bb$^+$	0.330	24.2
bb$^+$y^{2}eq/bb$^+$y^{2}eq	0.578	29.0
bb$^+$/sc^{4}sc^8	0.185	30.4
bb$^+$y^{2}eq/sc^{4}sc^8	0.289[a]	31.6
bb$^+$/bb$^+$y^{2}eq	0.474[a]	28.7
w^{a}bb^1/bb$^+$	0.215[a]	25.8
w^{a}bb^1/car bb	0.115[a]	30.7
In(1) sc^{8}bb^1/car bb	0.112[a]	29.8
car bb/sc^{4}sc^8	0.085	29.0
bb(UC03)/sc^{4}sc^8	0.075[a]	29.3
ywbb^{ds}/sc^{4}sc^8	0.102	26.2
XY$^L \cdot$Y^S bb/sc^{4}sc^8	0.09 [a, b]	30.1
XY$^L \cdot$Y^S bb/w^{a}bb^1	0.12 [a, b]	29.2
g^2 ty bb/w^a bb^1	0.10 [a, b]	31.6

[a] Estimates made from X chromosome contributions in different genotypic combinations.

[b] These figures represent the maximum estimates.

TABLE 2
RNA/DNA Ratios in Adult Males of Several Genotypes

Genotype	rRNA/DNA $\times$ 100 at saturation level	RNA/DNA
bb^+/bb^+	0.370	22.5
bb^+y^2eq/bb^+	0.474	26.6
sc^4sc^8/bb^+	0.185	25.2
bb^+/O	0.185	22.4
sc^4sc^8/Y^{bb} N2	0.097	23.3
w^{m4}/Y^{bb} Su-Var5	0.165	28.1

TABLE 3
DNA and RNA Content in Pupae 45–55 Hours Old of Various Genotypes

Genotype	DNA[a]	RNA[a]	RNA/DNA
bb^+/bb^+ (0.370)	0.39	9.5	24.4
bb^+/sc^4sc^8 (0.185)	0.31	8.6	27.7
bb^+y^2eq/bb^+y^2eq (0.580)	0.25	6.4	25.6
car bb/sc^4sc^8 (0.085)	0.34	7.6	22.4

[a] Values expressed in micrograms per pupa.

a subnormal amount of ribosomes, the unaffected tissues could have a normal or almost normal amount of ribosomes.

An insight into the above-mentioned possibilities might be obtained by considering RNA/DNA ratios at an intermediate stage of development. Such an analysis, on pupae 45–55 hours old (Table 3) again shows no differences.

It seems quite likely that similar results would be obtained at other stages of development since the retardation of development of bobbed flies is through all stages and is not confined to any particular one. We then studied the rates of rRNA synthesis in flies with different rDNA content.

The various tissues of an organism exhibit different rates of rRNA synthesis, yet most of them contain the same number of copies of the genes from which rRNA is transcribed (Ritossa *et al.*, 1966b; Brown and Weber, 1968; Mohan *et al.*, 1969). In principle, to study the biochemical effects of a genetic deletion, only those tissues should be chosen where the genes involved in the deletion are maximally transcribed. With regard to genes for rRNA in *Drosophila*, the best candidate among all tissues seems to be the ovaries. This tissue is one of those most actively engaged in metabolism in the adult form of *Drosophila*. Furthermore, after a rapid phase of growth, it reaches a plateau of activity which is characterized by an almost constant egg production.

In the experiments to be described, ovaries from mated females were examined. Seven-day-old females were used in every experiment; mating was always with X/O males. The purpose of this work was to compare rRNA synthesis in stocks of different rDNA content. It was mentioned earlier that most of the tissues contain the same amount of rDNA, but the case of oocytes is rather peculiar. The recent literature speaks in terms of amplification of rDNA in this tissue. This was shown to be the case in amphibians (Brown and Dawid, 1968; Evans and Birnstiel, 1968; Gall, 1968) and also in insects (Gall *et al.*, 1969; Lima de Faria *et al.*, 1969). Hence it was necessary to verify whether the rDNA content in ovaries of *Drosophila* is the same as that obtained from the entire adults. The results referring to such an analysis of 4 stocks used in this study are shown in Table 4, and show no major differences between the values obtained for adults and ovaries. Minor differences might be due to the type of DNA extraction followed in the two cases. To study the rate of rRNA synthesis we measured rRNA specific activity. In order to use this as a measure of rate of rRNA synthesis one has to know the amount of rRNA already present in the cells. Table 5 shows the RNA/DNA ratios obtained using ovaries of flies having different rDNA content. No difference was found and the specific activity can thus be used as a measure of net synthesis provided the incorporation is linear during the time in which the measurement is made and pools of precursors are the same. Labeling was made both *in vitro* and *in vivo*. The *in vitro* experiments were performed by incubating dissected ovaries, previously washed with Ringer solution, in the same Ringer solution

TABLE 4

Ribosomal DNA Content in Ovaries and Adult Female Flies of Four Genotypes

Genotype	rRNA/DNA $\times$ 100 at saturation in ovaries	rRNA/DNA $\times$ 100 at saturation in adults
bb$^+$/bb$^+$	0.317	0.330
bb$^+$y^{2}eq/bb$^+$y^{2}eq	0.541	0.578
bb$^+$/sc^{4}sc^8	0.207	0.185
car bb/sc^{4}sc^8	0.106	0.085

TABLE 5
RNA and DNA Content in Ovaries of Four Genotypes

Genotype	RNA[a]	DNA[a]	RNA/DNA
bb⁺/bb⁺ (0.370)	6.6	0.16	41.3
bb⁺y²eq/bb⁺y²eq (0.580)	5.9	0.13	44.2
bb⁺/sc⁴sc⁸ (0.185)	7.8	0.18	42.7
car bb/sc⁴sc⁸ (0.085)	7.2	0.15	46.3

[a] Values expressed in micrograms per pair of ovaries.

containing ^{3}H-labeled uridine. Figure 1 shows the time-course of incorporation of ^{3}H-labeled uridine in ovaries dissected from wild-type flies. The incorporation is linear for about 1 hour. The various RNA species were purified and fractionated on MAK columns. The problem of eventual differences in pools was bypassed by considering, in the different experiments, the specific activity of ribosomal RNA versus that of transfer RNA, the assumption being that tRNA and rRNA precursor pools are the same.

Figure 2 shows the specific activity of rRNA obtained after 30-minute pulses with tritiated uridine in ovaries of 6 genotypes carrying different amounts of rDNA. The rate of synthesis increases with the number of genes for rRNA up to a threshold, but no effect of the number of genes on the rate of rRNA synthesis is found afterward. tRNA specific activity is about the same in all cases (Table 6).

The rRNA labeled after 30 minutes is displaced from the OD on MAK columns indicating that most of the freshly synthesized rRNA is in the form of precursor molecules. That this is indeed the case was shown by sucrose density gradient analysis and polyacrylamide gel electrophoresis. The sedimentation value of this rRNA precursor molecule is 38 S, and its molecular weight, as calculated by polyacrylamide gel electrophoresis (Bishop et al., 1967), is 2.9×10^6 Daltons.

The net amount of rRNA synthesis in short pulses is less in the bobbed than the wild-type. If these observations are correct one has to explain how the same RNA/DNA ratios are obtained in bb as well as in wild-type individuals. One explanation could be that the conversion of ribosomal RNA precursors to mature rRNAs is different in bb and wild individuals. The in vitro studies on rRNA synthesis do not offer the possibility of studying the conversion of rRNA precursors to stable species of rRNA (Edström and Daneholt, 1967; Greenberg, 1967) even if the synthesis of these molecules occurs normally. We turned then to study the problem in vivo.

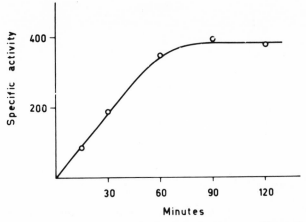

FIG. 1. Time course of rRNA synthesis *in vitro*. Ovaries were dissected out of wild-type females and incubated in Ringer solution containing ^{3}H-labeled uridine as described in Materials and Methods. RNAs were fractionated on MAK columns. Specific activities are expressed as counts per minute per microgram of rRNA. The reasons determining the flattening of rRNA specific activity in longer times were not studied; however, the incorporation is linear for about one hour.

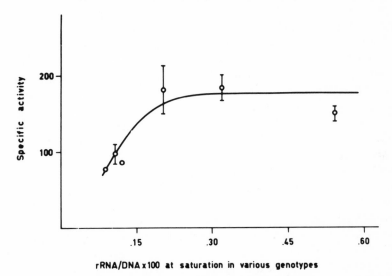

FIG. 2. Specific activity of rRNA after 30-minute pulse with ^{3}H-labeled uridine of ovaries dissected from females of various rDNA content. Rest as in Fig. 1.

49

TABLE 6
In Vitro RNA Synthesis in Ovaries of Various Genotypes[a]

Genotype	Specific activity (cpm/μg RNA)		
	rRNA	tRNA	rRNA/tRNA
bb^+/bb^+ (0.370)	185	49	3.7
bb^+y^2eq/bb^+y^2eq (0.580)	151	41	3.7
bb^+/sc^4sc^8 (0.185)	182	43	4.2
car bb/sc^4sc^8 (0.085)	97	52	1.9
$XY^L \cdot Y^S$ bb/w^abb^1 (0.120)	77	49	1.6
$XY^L \cdot Y^S$ bb/sc^4sc^8 (0.090)	75	46	1.6

[a] Flies were mated with X/O males for 7 days before experiment.

The *in vivo* studies on RNA synthesis have been carried out by feeding the flies on ^{32}P medium. Flies were removed at different intervals, and ovaries were dissected out from them. Again, to avoid differences in food composition, amounts eaten, and pool size of phosphorus, the results are expressed by comparing the ratio between the specific activities of rRNA and tRNA.

The results of an experiment where females of bb^+/bb^+ (0.370) and $XY^L \cdot Y^S$ bb/sc^4sc^8 (0.090) composition were employed are listed in Table 7. Both ovaries and carcass were used for RNA extractions, after feeding the flies for 4 hours. It is interesting to note that the differences are only in the ratio calculated on ovaries, not on carcass. Furthermore (although not mentioned in Table 7), the specific activity of RNA from carcass is nearly four times less than that from the ovaries. In the same table, ratios of another experiment using ovaries from bb^+/bb^+ (0.370) and car bb/sc^4sc^8 (0.085) genotypes are also presented. The flies were fed for 39 hours, and one can see that the differences diminish considerably with the increase of time.

Since there are no differences in the RNA synthesis of carcass between the wild and bobbed genotypes and, in addition, the ovaries contain nearly 65% of the total RNA present in the 7-day-old fly, the experiments can also be performed using whole flies. Such studies have been made using another bobbed genotype, $XY^L \cdot Y^S$ bb/w^abb^1 (0.120) along with wild type. The results are shown in Table 8. The bobbed mutant shows greater differences in shorter periods in rRNA synthesis and the differences become less and less conspicuous with the passage of time.

Since *bobbed* mutants synthesize rRNA at a lower rate in the

TABLE 7
In Vivo Synthesis of RNA in Ovaries and Carcass of Various Genotypes

Genotype	Ratio of specific activity between rRNA and tRNA	
	In ovaries	In carcass
	After 4 hours of feeding	
bb$^+$/bb$^+$ (0.370)	1.03	0.77
XY^L · Y^S bb/sc^{4}sc^8 (0.090)	0.86	0.75
	After 39 hours of feeding	
bb$^+$/bb$^+$ (0.370)	0.77	—
car bb/sc^{4}sc^8 (0.085)	0.65	—

TABLE 8
In Vivo Synthesis of RNA in Two Genotypes[a]

Genotype	Ratio of specific activity between rRNA and tRNA		
	Time in hours after feeding		
	2	4	8
bb$^+$/bb$^+$ (0.370)	1.51	1.02	0.75
XY^L · Y^S bb/w^{a}bb^1 (0.120)	0.81	0.74	0.64

[a] Whole flies, mated with X/O males, were employed.

TABLE 9
RNA and DNA Content of Eggs Laid by Females of Various Genotypes[a]

Genotype	RNA	DNA × 10^3	RNA/DNA
bb$^+$/bb$^+$ (0.370)	0.14	0.91	153
bb$^+$/sc^{4}sc^8 (0.185)	0.12	0.71	166
bb$^+$y^{2}eq/bb$^+$y^{2}eq (0.580)	0.13	0.93	136
car bb/sc^{4}sc^8 (0.085)	0.14	1.02	143
XY^L · Y^S bb/w^{a}bb^1 (0.120)	0.11	—	—
XY^L · Y^S bb/sc^{4}sc^8 (0.090)	0.13	—	—

[a] Females were mated with X/O males. Values are expressed in micrograms per egg.

TABLE 10
EGG LAYING RATE FOR 6 GENOTYPES OF *DROSOPHILA MELANOGASTER* AT DIFFERENT AGES

Age in hours	Number of eggs per female per hour in different genotypes					
	bb^-/bb^- (0.370)	bb^-y^2eq/bb^-y^2eq (0.580)	bb^-/sc^4sc^8 (0.185)	car bb/sc^4sc^8 (0.085)	$XY^L \cdot Y^S bb/w^a bb^l$ (0.120)	$XY^L \cdot Y^S bb/sc^4sc^8$ (0.090)
0	—	—	—	—	—	—
23	—	—	0.32	—	—	—
35	0.26	NO[a]	NO	0.097	—	—
48	1.49	1.32	1.79	0.86	0.23	0.26
59	NO	1.86	2.72	1.65	1.14	1.20
73	2.09	1.58	1.91	2.16	1.17	0.86
83	2.96	2.24	3.20	2.31	NO	NO
96	2.01	NO	NO	1.62	1.54	1.39
101	NO	2.50	3.05	NO	—	NO
108	2.65	NO	NO	2.14	2.16	0.97
123	2.21	2.44	3.61	2.44	1.78	1.02
146	2.75	2.99	2.98	2.63	2.06	1.58
158	2.65	NO	NO	NO	1.82	1.32
170	2.60	2.34	2.70	2.12	2.24	1.32
185	3.01	NO	NO	NO	0.96	1.47
197	2.30	2.10	3.15	2.28	1.35	1.27
214	2.70	NO	NO	NO	1.13	1.49
221	2.82	2.44	3.20	2.72		
251	3.59	2.70	3.41	3.06	Mean[c] 1.62	1.27
276	2.76	2.21	2.93	3.15		
298	NO	1.96	2.85	NO		
305	2.11	NO	NO	3.23		
321	NO	2.13	3.12	NO		
327	1.80	NO	NO	2.99		
350	2.13	1.73	2.60	2.17		
373	1.97	1.99	2.96	2.23		
397	1.04	NO	NO	2.06		
Mean[b]	2.51	2.24	2.98	2.48		
SD	0.465	0.360	0.389	0.446		

[a] Not observed.

[b] Calculations done only in the age period from 73 to 373 hours for first 4 genotypes.

[c] Calculation done only in the age period of 73 to 214 hours.

ovaries, both *in vitro* and *in vivo*, than the wild-type, we looked for the consequences of such a phenomenon. Since the RNA/DNA ratio is the same in the two cases, possible consequences of the reduced rate of synthesis in bb ovaries was expected to influence either (a) the rRNA content of eggs, or (b) the total number of eggs the bb flies are capable of producing. Table 9 shows that the RNA/DNA ratio in the eggs produced by flies with different rDNA content is the same.

Table 10 shows the egg-laying ability of females having different numbers of genes for ribosomal RNA. No difference is observed between females with nearly 600, 360, or 180 genes. Among the females showing the bobbed phenotype, car bb/sc^{4}sc^8 (0.085) shows a difference in the early stages but no significant difference in late periods. Other bobbed females, XY$^L \cdot$ Y^S bb/sc^{4}sc^8 (0.090) and XY$^L \cdot$ Y^S bb/w^{a}bb^1 (0.120), which show a more pronounced phenotype (the rDNA content of the XY$^L \cdot$ X^S bb chromosome is a maximum estimate) have a strongly reduced egg-laying ability all through the deposition time. This is also shown in Fig. 3. Bobbed females can be produced which are even stronger in bb phenotype than the one presented here. We analyzed females of composition g^2 ty bb/sc^{4}sc^8. A direct analysis of the rDNA content in these females is extremely difficult. Indeed very few of them reach the adult stage. An estimate of their rDNA content can be obtained, however, by analyzing the rDNA content carried by the g^2 ty bb chromosome in other combinations. (The sc^{4}sc^8 chromosome carries no rDNA) The first data indicate that the rDNA content of such females should be around 0.07%). These females produce such a limited number of eggs that tabulation has been impossible.

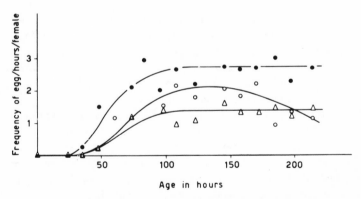

Fig. 3. Curves for egg-laying rate of three genotypes are shown over a period of time. ●, bb$^+$/bb$^+$ (0.360); O, XY$^L \cdot$ Y^{S}bb/w^{a}bb^1 (0.120); △, XY$^L \cdot$ Y^{S}bb/sc^{4}sc^8 (0.090). Rest as in Table 10.

In this work we analyzed (a) whether every gene for rRNA present in a cell is transcribed or not, that is, whether there are systems of regulation for the synthesis of rRNA, and (b) how the bobbed phenotype comes about when the cells carry a low rDNA content.

Regulation of rRNA synthesis. Brown and Gurdon (1964) had shown that *Xenopus* embryos having one-half the normal rDNA content (Birnstiel *et al.*, 1966) were able to synthesize as much rRNA as embryos carrying a normal rDNA content. Other reports, instead, indicate that the number of nucleolus organizers, and hence probably the number of genes for rRNA (Ritossa *et al.*, 1966a) and total RNA content are positively correlated both in maize and wheat (Lin, 1955; Crosby-Longwell and Svihla, 1960).

In *Drosophila*, the rDNA content *versus* phenotype can thus be summarized:

Genetic composition	(%) rDNA	Phenotype
bb/0 (or bb/bb)	Less than 0.130	bb
bb$^+$/0 (or bb/bb)	About 0.130	+
bb$^+$/bb$^+$	From about 0.260 to 0.600	+

One might expect that, under certain conditions (e.g., template limitation) variations of the rate of RNA synthesis might not be the consequence of regulatory mechanisms but only represent kinetic effects.

Since we have seen that the bb phenotype is due to a reduced rate of rRNA synthesis, we can conclude that the rate of rRNA synthesis in *Drosophila* cannot be increased, if at all, more than twice. Indeed, if the rate of transcription of rDNA could be enormously increased, bb mutations would be unrecognized, or only very strong rDNA deletions could be revealed phenotypically.

We cannot state at present whether the transcription rate of rDNA is constant irrespective of the rDNA content, or varies within a $2\times$ ratio. However, the existence of mechanisms able to limit rDNA transcription will be revealed by the rate of rRNA synthesis in stocks having higher rDNA content than an individual with two bb$^+$ iso-alleles, only one of which is enough to give a wild phenotype.

Schultz and Travaglini (1965) examined the ribosomal content of eggs produced by females carrying 2 and 3 nucleolus organizers and found no difference between them. Keifer (1968) examined the ribosome content of flies having 1, 2, and 3 nucleolus organizers and also found no difference.

We know, however, that there is no necessary correlation between the number of nucleolus organizers and rDNA content (Ritossa and Scala, 1969).

Our 30-minute pulses monitored almost exclusively the synthesis of a 38 S precursor to ribosomal RNA, while the long-term labelings monitored predominantly mature rRNAs. Since both the *in vitro* (short term) and the *in vivo* (long term) experiments gave similar results, we can conclude that the rate of transcription of rRNA in *Drosophila* ovaries is dependent on the rDNA content up to a certain threshold which corresponds to the number of genes sufficient to give a wild phenotype.

Ovaries, whose cells carry as many as 600 genes for rRNA, show a rate of rRNA synthesis which is similar to that found in ovaries whose cells carry about 180 such genes (bb$^+$/O) and about 360 genes (bb$^+$/ bb$^+$).

We can hence conclude that not all rDNA in every cell is necessarily used at the same time. Mechanisms have to exist to regulate the amount of rRNA which has to be synthesized in a certain period of time.

Genesis of the bobbed phenotype. How an rDNA partial deletion leads to a bobbed phenotype was studied using ovaries, the most active tissue of adult females.

Both *in vitro* and *in vivo* experiments clearly showed that the specific activity of rRNA (and hence the rate of rRNA synthesis) is lower in ovaries of bobbed flies. As a consequence of this we observed both a retardation in the appearance of the egg-laying ability and a strongly reduced egg-laying rate with time.

In one case (car bb/sc^{4}sc^8) (0.085) we could observe a reduced egg laying rate only in the first days. This might be due merely to experimental errors or to some unconsidered parameter. The correlation between a low rRNA synthesis and a reduced egg-laying ability is rather striking in the other bb isoallele studied (XY^L·Y^S bb) (0.090), and using a stronger bb isoallele (g^2 ty bb) (0.070) on which we could not make studies of rRNA synthesis, we found almost complete sterility. Schultz and Travaglini (personal communication) also found an inverse proportionality between the intensity of the bobbed phenotype and egg-laying ability.

Since the rRNA content per egg is not affected by a low rDNA content, mechanisms must exist such that, until an egg accumulates a normal, or almost normal amount of rRNA, it is not laid. Furthermore, bobbed females require a longer time to start deposition. Since we know that ovaries of bb females show a reduced rate of rRNA synthesis while the amount of rRNA per unit DNA in the mature ovary is the same as that of wild-type flies, it is most likely that this retardation allows the accumulation of ribosomes to make an efficient machinery for egg production. In light of this, it is not surprising to find that in adult bobbed flies and pupae the RNA/DNA ratio is the same as that found in phenotypically wild individuals.

The bb flies have been found to have a longer developmental time. A reduced rate of rRNA synthesis can be compensated for by an increase in the time of development. Were this absolutely correct, one would expect that no progress in development would occur until *every* cell of the organism has reached the ribosome content characterizing the corresponding wild-type cells. In this case, one would expect that the only trait influenced by the bb mutation would be developmental time; one might obviously expect characters which develop after hatching time (fertility, for instance) to be influenced. Obviously this is not the case.

We propose then that a normal ribosome content has to be achieved only in tissues most relevant for the progress of development. Other, less relevant, tissues or cells can develop even with subnormal amounts of ribosomes; they will lead to the mutated traits.

SUMMARY

The rate of ribosomal RNA (rRNA) synthesis was studied both *in vitro* and *in vivo* using ovaries from *Drosophila melanogaster* stocks carrying different amounts of DNA complementary to ribosomal RNA (rDNA). It was shown that the rate of rRNA synthesis is reduced in flies exhibiting the bobbed phenotype and having an rDNA content lower than 0.130%. These flies have been shown to lay eggs at a reduced rate but of normal RNA content. Ovaries from phenotypically wild flies, with rDNA contents of 0.180%, 0.370%, and 0.580% showed instead about the same rate of rRNA synthesis. Females with such rDNA contents lay eggs of the same RNA content and at the same rate. These data allowed us to postulate the existence of regulatory mechanisms for rRNA synthesis. In spite of the reduced rate of synthesis in certain tissues, bobbed flies (both males and females) have been shown to have the same RNA/DNA ratio as phenotypically wild flies with different rDNA contents. The same results have been obtained using an intermediate stage (pupae). It is known that bobbed flies exhibit a delayed development. Probably this allows for a normalization of the rRNA content. The genesis of the bobbed phenotype is discussed on the basis of the results obtained.

We are grateful to Dr. Arthur Soller for his help in the revision of the manuscript.

REFERENCES

BISHOP, D. H. L., CLAYBROOK, J. R., and SPIEGELMAN, S. (1967). Electrophoretic separation of viral nucleic acids on polyacrylamide gels. *J. Mol. Biol.* **26**, 373–387.

BIRNSTIEL, M. L., WALLACE, H., SIRLIN, J. L., and FISHBERG, M. (1966). Localization of the ribosomal DNA complements in nucleolar organizer region of *Xenopus laevis*. *Nat. Cancer Inst. Monogr.* **23**, 431–444.

BROWN, D. D. and DAWID, I. B. (1968). Specific gene amplification in oocytes. *Science* **160**, 272–280.

BROWN, D. D. and GURDON, J. B. (1964). Absence of ribosomal RNA synthesis in the anucleolate mutant of *Xenopus laevis*. *Proc. Nat. Acad. Sci. U.S.* **51**, 139–146.

BROWN, D. D., and WEBER, C. S. (1968). Gene linkage by RNA-DNA hybridization. I. Unique DNA sequences homologous to 4S RNA, 5S RNA and ribosomal RNA. *J. Mol. Biol.* **34**, 661–680.

BURTON, K. (1956). A study of the conditions and mechanism of the diphenylamine reaction for the colorimetric estimation of deoxyribonucleic acid. *Biochem. J.* **62**, 315–323.

CROSBY-LONGWELL, A., and SVIHLA, G. (1960). Specific chromosomal control of the nucleolus and of the cytoplasm in wheat. *Exp. Cell Res.* **20**, 294–312.

EDSTROM, J. E., and DANEHOLT, B. (1967). Sedimentation properties of the newly synthesized RNA from isolated nuclear components of *Chironomus tentans* salivary gland cells. *J. Mol. Biol.* **28**, 331–343.

EVANS, D., and BIRNSTIEL, M. L. (1968). Localization of amplified ribosomal DNA in the oocyte of *Xenopus laevis*. *Biochim. Biophys. Acta* **166**, 274–276.

GALL, J. G. (1968). Differential synthesis of the genes for ribosomal RNA during amphibian oogenesis. *Proc. Nat. Acad. Sci. U.S.* **60**, 553–560.

GALL, J. G., McGREGOR, H. C., and KIDSTON, M. E. (1969). Gene amplification in the oocytes of Dytiscus water beetles. *Chromosoma* **26**, 169–187.

GILLESPIE, D., and SPIEGELMAN, S. (1965). A quantitative assay for DNA-RNA hybrids with DNA immobilized on a membrane. *J. Mol. Biol.* **12**, 829–842.

GREENBERG, J. R. (1967). Sedimentation studies on *Drosophila virilis* salivary gland RNA. *J. Cell Biol.* **35**, 49A.

KIEFER, B. I. (1968). Dosage regulation of ribosomal DNA in *Drosophila melanogaster*. *Proc. Nat. Acad. Sci. U.S.* **61**, 85–89.

LIMA DE FARIA, A., BIRNSTIEL, M. L., and JAWOROSKA, H. (1969). Amplification of ribocistrons in the heterochromatin of *Acheta*. *Genetics*, Suppl., **61**, 145–159.

Lin, M. (1955). Chromosomal control of nuclear composition in maize. *Chromosoma* **7,** 340–370.

Mohan, J., Dunn, A., and Casola, L. (1969). Ribosomal DNA in the rat. *Nature* **223,** 295–296.

Munro, H. N., and Fleck, A. (1966). The determination of nucleic acids. *In* "Methods of Biochemical Analysis," (D. Glick, ed.), Wiley (Interscience), New York. Vol. 14, pp. 113–176.

Ritossa, F. M., and Scala, G. (1969). Equilibrium variations in the redundancy of rDNA in *Drosophila melanogaster*. *Genetics* **61,** Suppl., 305–317.

Ritossa, F. M., Atwood, K. C., and Spiegelman, S. (1966a). A molecular explanation of the bobbed mutants of Drosophila as partial deficiencies of "ribosomal" DNA. *Genetics* **54,** 819–834.

Ritossa, F. M., Atwood, K. C., Lindsley, D. L., and Spiegelman, S. (1966b). On the chromosomal distribution of DNA complementary to ribosomal and soluble RNA. *Nat. Cancer Inst. Monogr.* **23,** 449–471.

Schmidt, G., and Thannhauser, S. J. (1945). A method for the determination of desoxyribonucleic acid, ribonucleic acid and phosphoproteins in animal tissues. *J. Biol. Chem.* **161,** 83–89.

Schultz, J., and Travaglini, E. (1965). Evidence for homeostatic control of ribosomal content in *Drosophila melanogaster* eggs. *Genetics* **52,** 473.

Cytological Localization of DNA Complementary to Ribosomal RNA in Polytene Chromosomes of *Diptera**

Mary Lou Pardue, Susan A. Gerbi, Ronald A. Eckhardt, and Joseph G. Gall

Introduction

Numerous studies indicate that the genes coding for ribosomal RNA[1] are located in or near the nucleolus organizer (Vincent and Miller, 1966). The most conclusive evidence comes from the molecular hybridization experiments of Ritossa and Spiegelman (1965) and Wallace and Birnstiel (1966), who have used combined genetic and biochemical means to localize the rDNA in *Drosophila melanogaster* and *Xenopus laevis* respectively.

Recently techniques have been described for the hybridization of nucleic acids in cytological preparations (Gall, 1969; Gall and Pardue,

* Supported by funds from Public Health Service grants GM 12427, GM 397, and 4 FOl GM 33363 from the National Institute of General Medical Sciences and 5 TOl HD 32 from the National Institute of Child Health and Human Development.

1 Abbreviations. rRNA: ribosomal RNA; rDNA: the DNA which codes for ribosomal RNA; SSC: 0.15 M NaCl, 0.015 M Na citrate, pH 7.0; EDTA: ethylene diamine tetraacetate; tris: tris-(hydroxymethyl)-amino methane; SDS: sodium dodecyl sulfate; ATP, CTP, GTP, UTP: the triphosphates of adenosine, cytidine, guanosine, and uridine; TCA: trichloroacetic acid.

1969; Pardue and Gall, 1969; John, Birnstiel, and Jones, 1969; Buon-giorno-Nardelli and Amaldi, 1969). With these methods it is possible for the first time to study the cytological localization of various genes on the basis of their nucleotide sequences. We now report experiments in which radioactive rRNA has been hybridized with the salivary gland nuclei of three Dipterans: *Drosophila hydei, Sciara coprophila,* and *Rhynchosciara hollaenderi.* A preliminary account of these findings has appeared in abstract form (Pardue, 1969).

Materials and Methods

1. Cytological Hybridization

Slides were made from the salivary glands of third instar larvae of *Drosophila hydei* and fourth instar larvae of *Rhynchosciara* and *Sciara coprophila.* The glands were removed in Ringer's solution. *Rhynchosciara* and *Sciara* salivary glands were fixed for a few minutes in 3:1 ethanol-acetic acid and squashed under a coverslip in 45% acetic acid. Glands from *D. hydei* were fixed in 45% acetic acid before being squashed under a coverslip in the same solution. The slides were frozen on dry ice, the coverslips were removed from the preparations with a razor blade, and the slides were quickly placed in 95% ethanol (Conger and Fairchild, 1953). After this step the slides were air dried.

The preparations were treated at room temperature for 2 hours with pancreatic RNase (100 μg/ml in 2 X SSC). This treatment was followed by three washes in 2 X SSC, one in 70% ethanol, and one in 95% ethanol before air drying. Dry slides were dipped into 0.5% agar at 60° C and then placed in 0.07N NaOH at room temperature for 2 min or more to denature the DNA of the preparation. After the denaturation step, the slides were washed twice in 70% ethanol and twice in 95% ethanol before drying. In more recent experiments we have omitted the agar coating step.

For hybridization 50—200 μl of 6 X SSC containing the radioactive RNA was placed over each salivary gland preparation and covered with a coverslip. *Xenopus* rRNA was used at 2 μg/ml; the complementary RNA transcribed from *Xenopus* rDNA was used at 4×10^6 cpm/ml. The slides were placed in moist chambers made from plastic Petri plates and were incubated at 66° C for 12—15 hours. After the hybridization step, coverslips were removed by dipping the slides in 6 X SSC. Slides were washed in 6 X SSC and then treated with pancreatic RNase (20 μg/ml in 2 X SSC) for one hour at 37° C. RNase treatment was followed by washing in 6 X SSC, 70% ethanol, and 95% ethanol. The slides were then air dried.

Dry slides were coated with Kodak NTB-2 emulsion diluted 1:1 with water and were stored in light-tight boxes at 4° C. The preparations were developed for 2 minutes in Kodak D-19, rinsed in 2% acetic acid, and fixed for 3 minutes in Kodak Fixer. After several washes in distilled water, slides were stained with Giemsa and mounted under a coverslip.

2. Preparation of Radioactive Ribosomal RNA

It has been shown that the rRNA of several eukaryotes will cross-hybridize with the DNA of other eukaryotic species (Brown, Weber, and Sinclair, 1967). This fact made it possible for us to use rRNA from the toad, *Xenopus laevis,* in our experiments on Dipteran salivary glands. Tritiated rRNA of high specific activity was obtained by growing a culture of *Xenopus* kidney cells for 9 days in a medium containing uridine-H³ (50 µC/ml, 20 C/mM). The cells were lysed at 4° C with sodium lauroyl sarcosinate (1% Sarkosyl Geigy in 0.1 M Na acetate, 4 µg/ml polyvinyl sulfonate, pH 5.0), and protein was extracted with an equal volume of cold phenol. The aqueous phase was brought to 0.1 M NaCl and the nucleic acids were precipitated by the addition of two volumes of 95% ethanol. The precipitate was dissolved in buffer containing 0.01 M Na acetate, 0.001 M $MgCl_2$, 1 µg/ml polyvinyl sulfonate, pH 5.0, and treated with DNase I (5 µg/ml) at room temperature for 10 minutes. After a second ethanol precipitation, the RNA was dissolved in 0.01 M Na acetate, 1 µg/ml polyvinyl sulfonate, pH 5.0, and centrifuged on a 5—20% (w/v) sucrose gradient at 4° C. The 18 S and 28 S peaks were combined for use in our experiments. The combined rRNA had a specific activity of 500,000 cpm/µg as determined by scintillation counting of known amounts of RNA on nitrocellulose filters (efficiency of counting about 10%).

3. Preparation of Complementary RNA by in vitro transcription

Highly radioactive RNA complementary to the satellite DNA of *Xenopus* was prepared by *in vitro* transcription using *Escherichia coli* polymerase. DNA was extracted from the ovaries of animals a few weeks past metamorphosis. At that stage the ovary contains many oocytes in which there have been extra replications of the satellite DNA (Gall, 1968). DNA was extracted as previously described (Pardue and Gall, 1969) and centrifuged to equilibrium on a CsCl density gradient. The gradient fractions which contained the satellite DNA were pooled and the DNA was recovered by ethanol precipitation.

RNA polymerase was prepared from frozen cells of *E. coli*, strain A-19, by the procedure described by Burgess (Burgess, Travers, Dunn, and Bautz, 1969). After the DEAE-cellulose column chromatography, fractions which contained the enzyme were pooled. The enzyme was precipitated with 50% $(NH_4)_2SO_4$ and redissolved in the storage buffer. The polymerase was used in our transcription experiments without further purification.

The transcription was carried out in 0.25 ml of buffer containing 4 units of enzyme, 5 µg satellite DNA, 60 mµ moles GTP, and 100 µC each of ATP-H³ (24.5 C/mM), CTP-H³ (13.5 C/mM), and UTP-H³ (14.8 C/mM). The buffer composition was 0.04 M tris pH 7.9, 0.15 M KCl, 0.0046 M $MgCl_2$, 0.002 M $MnCl_2$, 7×10^{-5} M EDTA, and 0.0058 M β-mercaptoethanol. The reaction mixture was incubated for $1^1/_2$ hours at 37° C. DNase I (Worthington DPFF, 20 µg in 0.75 ml of 0.04 M tris, pH 7.8) was then added and the mixture held at room temperature for 20 minutes. Non-radioactive *E. coli* rRNA (80 µg of 16 S and 80 µg of 23 S in 1 ml 0.1 X SSC) was added as carrier, followed by 200 µl 5% SDS and 2 ml phenol. The mixture was held at 4° C overnight. After centrifugation, the phenol layer was re-extracted with a small amount of water. The aqueous layers from the two centrifugations were pooled, heated to 85° C for 3 minutes, and placed on a Sephadex G-50 column (D. Brown, personal communication). The column was eluted with water and the fractions containing TCA insoluble counts were pooled, heated to 85° C for 3 min, and put through a nitrocellulose filter (0.45 µ pore size). Approximately 2 µg of complementary polynucleotide was produced. It had a calculated specific activity of 7×10^7 dpm/µg (about 7×10^6 cpm/µg as counted on nitrocellulose filters).

4. Filter Hybridization

The *Rhynchosciara* DNA used in our filter hybridizations was extracted from adult females which had been stored at -20° C. The animals were briefly homogenized at 4° C in a buffer consisting of 0.05 M tris, 0.025 M KCl, 0.005 M Mg acetate, and 0.35 M sucrose, pH 7.6. The homogenate was centrifuged at $5,000 \times g$ for 10 minutes. The resulting pellet, which contained the nuclei, was resuspended in a solution of 0.5% Sarkosyl Geigy, 100 µg/ml self-digested pronase, 0.1 M EDTA, and 0.05 M tris, pH 8.4. After a 2 hour incubation at 37° C the mixture was extracted twice with water-saturated phenol. The aqueous layer was then brought to 0.1 M NaCl and the nucleic acids were precipitated with 2 volumes of 95% ethanol. The precipitate was redissolved in 0.1 X SSC and treated at 37° C for two hours with 100 µg/ml pancreatic RNase, 330 units/ml T_1 RNase, and 250 µg/ml α-amylase. Pronase (100 µg/ml) was added and the incubation continued for one more hour.

In the final purification step the DNA was centrifuged to equilibrium in a CsCl density gradient. The centrifugation was for 20 hours at 42,000 rpm in a Spinco Ti-50 head. Ten-drop fractions were collected from the bottom of the centrifuge tube. After the optical density had been measured, each fraction was denatured with 0.1 N NaOH, neutralized, and then diluted with 6 X SSC. The samples were applied to nitrocellulose membrane filters and the hybridization was carried out as described by Gillespie and Spiegelman (1965). The filters were incubated with 2 µg/ml *Xenopus* rRNA in 6 X SSC for 12 hours at 66° C. Competition experiments included a 50-fold excess of non-radioactive *Rhynchosciara* or *Xenopus* rRNA during the hybridization step.

Results

1. Drosophila hydei

The presence of DNA in or on the nucleoli of Dipteran salivary gland nuclei has been reported many times since the original observations of Heitz and Bauer on *Bibio* (1933). In several species of *Drosophila*, including *D. hydei*, the intranucleolar DNA occurs in characteristic morphological patterns (Kaufmann, 1938; Swift, 1962; Barr and Plaut, 1966; Olvera, 1969; Rodman, 1969). In *D. hydei* salivary gland nuclei we found that radioactive *Xenopus* rRNA hybridized specifically with the DNA present within the nucleolus (Figs. 1—4). In our Giemsa stained preparations the nucleolar DNA typically occurred as one or more round masses surrounded by smaller strands and dots, in essentially the pattern described by Barr and Plaut (1966). In slide hybrids the silver grains of the autoradiographs showed the same distribution as did the DNA of the nucleolus. Occasionally it was possible to see a labeled DNA strand running out of the nucleolus toward a chromosome, but in most cases the nucleolus seemed to be completely free of the chromosomes. Undoubtedly the pressure applied during specimen preparation tended to separate the nucleolus from its original association with the chromocenter. We have not seen hybridization of rRNA to any band within the chromosomes.

2. Rhynchosciara hollaenderi

The arrangement of the nucleolar material in the salivary gland nuclei of *Rhynchosciara* is quite different from that of *D. hydei*. In *Rhynchosciara* salivary glands the end of the X chromosome bears a prominent

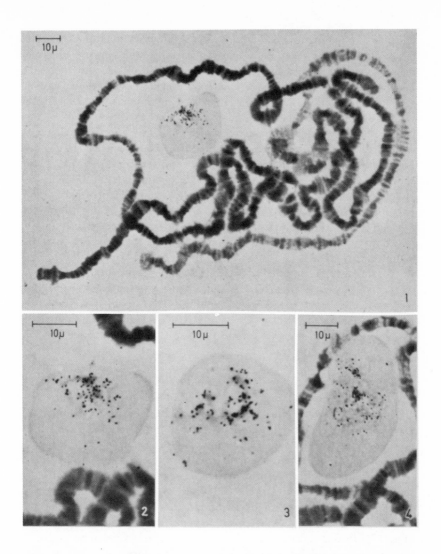

Figs. 1—4. Autoradiographs of a salivary gland preparation from a larva of *Droso-phila hydei* showing hybridization of radioactive ribosomal RNA to the DNA within the nucleolus. The DNA of the preparation was denatured with 0.07 N NaOH for 20 minutes and then hybridized with radioactive *Xenopus* ribosomal RNA for 14 hours. The RNA had a specific activity of 5×10^5 cpm/μg. The slide was stained with Giemsa. Exposure 94 days. Fig. 1. Entire chromosome set of *D. hydei*. Ribosomal cistrons are detected only within the body of the nucleolus. In this preparation the nucleolus is detached from the chromocenter. $\times 650$. Fig. 2. Enlargement of the nucleolus shown in Figure 1. The distribution of the hybridized RNA follows the distribution of the nucleolar DNA. $\times 1200$. Fig. 3. Free nucleolus. $\times 1500$.

Fig. 4. Nucleolus. $\times 880$

nucleolus in young larvae. During larval life this nucleolus tends to break up, and by late fourth instar variable amounts of nucleolar material are present on the X (Breuer and Pavan, 1955; Mattingly and Parker, 1968). Salivary gland nuclei of sciarid flies contain a second type of nucleolar material referred to as "micronucleoli". In salivary squashes from older larvae the micronucleoli occur as numerous smaller bodies scattered among the chromosomes, often in close association with certain bands (Swift, 1962; Pavan, 1965; Gabrusewycz-Garcia and Kleinfeld, 1966). The micronucleoli of several sciarid species have been shown to contain DNA (da Cunha, Morgante, and Garrido, 1969) and to have a fine structure resembling that of the true nucleolus (Swift, 1962; Jacob and Sirlin, 1963). In both *Rhynchosciara* and *Sciara* the micronucleoli contain Feulgen-positive granules and can incorporate thymidine (Eckhardt, unpublished observations; Gabrusewycz-Garcia, personal communication).

We have done cytological hybridization experiments on *Rhynchosciara* using both *Xenopus* rRNA and complementary RNA transcribed enzymatically from the *Xenopus* satellite DNA (which contains the ribosomal cistrons). The two types of RNA labeled the same regions in *Rhynchosciara* salivary glands, and the results described below apply to both types of experiment.

In *Rhynchosciara* salivary gland preparations we found that rRNA annealed invariably to the heterochromatic ends of the X and C chromosomes (see maps of Mattingly and Parker, 1968) and to the micronucleoli (Figs. 5—12). Silver grains were also found frequently over the heterochromatic tip of the B chromosome. The hybridization on the end of the X was clearly in the region of the nucleolus organizer (Figs. 7—9). In some preparations where a portion of the nucleolus was still attached to the X, we could see hybridized DNA splayed out into the nucleolar material (Fig. 8). The end of the X is a difficult region to analyze cytologically. Apparently there are two major masses of darkly stained material, but these are almost always paired in a confusing fold-back (Fig. 13). In our autoradiographs the silver grains tended to cluster in one or two groups and did not cover the entire darkly stained area.

Hybridization on the end of the C chromosome occurred as regularly as that on the end of the X (Figs. 9 and 10). No nucleolus has been reported on the C chromosome and we have not seen any nucleolar fragments associated with this region in late fourth instar larvae. However, we have not investigated the cytology of this region in younger animals. On the basis of the hybridization experiments we conclude that rDNA is regularly located in region 14 of the C chromosome.

Hybridization on the heterochromatic end of the B chromosome was variable, and in one series of slides was almost entirely lacking. Because

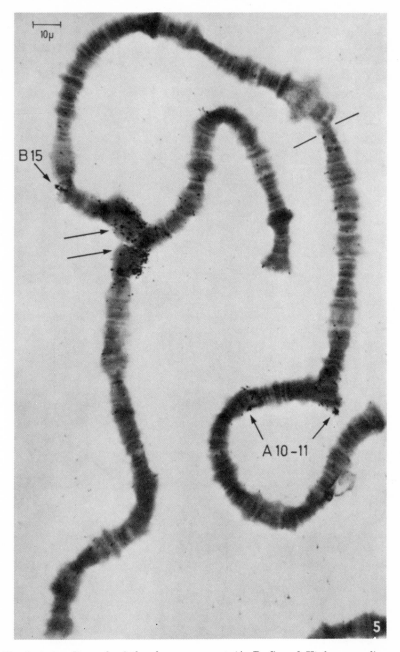

Fig. 5. Autoradiograph of the chromosome set (A, B, C, and X) from a salivary gland nucleus of *Rhynchosciara* hybridized with complementary RNA transcribed *in vitro* from the *Xenopus* satellite DNA. This satellite contains the ribosomal cistrons. The slide illustrates the association of heterochromatic regions of non-

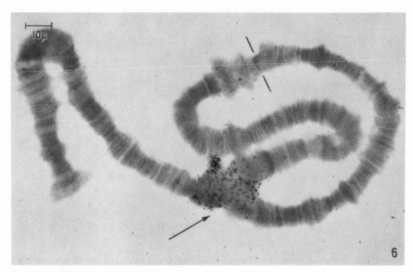

Fig. 6. Autoradiograph of chromosomes B, C, and X from a salivary gland nucleus of *Rhynchosciara*. Complementary RNA transcribed *in vitro* from the *Xenopus* satellite DNA (which contains the ribosomal cistrons) is hybridized to the heterochromatic ends of the three chromosomes (arrow). The conditions of hybridization are described in Figure 5. Terminal association of B and C chromosomes marked by a bar. Exposure 10 days. ×720

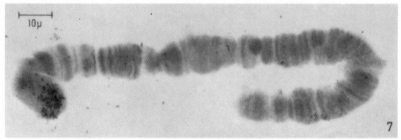

Fig. 7. The X chromosome from a salivary gland nucleus of *Rhynchosciara* showing hybridization of radioactive complementary RNA to the DNA of the heterochromatic end. This is the region which bears the nucleolus in earlier stages. The complementary RNA and the conditions of hybridization are the same as described in Fig. 5. Exposure 10 days. ×900

homologous chromosomes which is frequently seen in Dipteran salivary glands. The complementary RNA is hybridized to the region containing the associated heterochromatic ends of the X, the C, and the B chromosomes (double arrow) and to micronucleoli attached to regions A 10, A 11, and B 15 (single arrows). These are the same regions which bind *Xenopus* ribosomal RNA. Terminal association of A and B chromosomes marked by a bar. The DNA of this slide was denatured with 0.07 N NaOH for 2 minutes and then hybridized with complementary RNA for 10 hours. The RNA had a calculated specific activity of 7×10^7 dpm/μg. Slide stained with Giemsa. Exposure 10 days. ×770

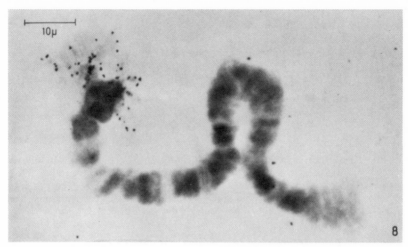

Fig. 8. The X chromosome from a salivary gland nucleus after hybridization with radioactive *Xenopus* ribosomal RNA. Strands of hybridized DNA radiate out into the portions of the nucleolus which still remain attached to this chromosome. The DNA of this chromosome was denatured for 3 minutes with 0.07 N NaOH and hybridized for 14 hours. The RNA had a specific activity of 5×10^5 cpm/µg. Giemsa stain. Exposure 21 days. $\times 1400$

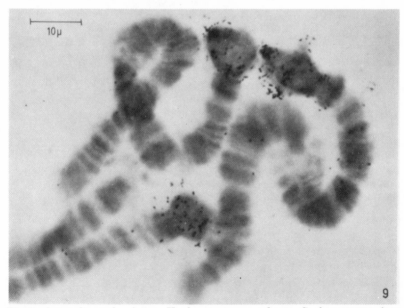

Fig. 9. Autoradiograph of two X chromosomes and one C chromosome from *Rhynchosciara* showing hybridization of radioactive *Xenopus* ribosomal RNA to the DNA of the heterochromatic ends. Conditions of hybridization were as described in Fig. 8. Exposure 80 days. $\times 1400$

67

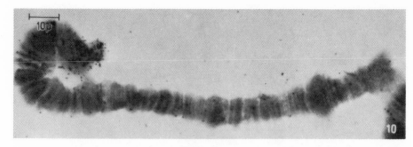

Fig. 10. The C chromosome from a *Rhynchosciara* salivary gland nucleus hybridized with radioactive complementary RNA transcribed from the *Xenopus* satellite DNA. This satellite contains the ribosomal cistrons. The RNA has bound to the same region which binds *Xenopus* ribosomal RNA. Conditions of hybridization as described in Fig. 5. Exposure 10 days. ×810

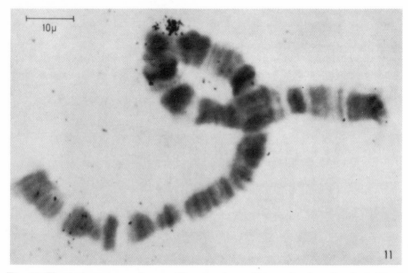

Fig. 11. The A chromosome from a *Rhynchosciara* salivary gland nucleus hybridized with radioactive *Xenopus* ribosomal RNA. The ribosomal RNA has bound to the DNA of a large micronucleolus lying beside region A 11. The conditions of hybridization are those described in Fig. 8. Exposure 89 days. ×1300

of the tight associations which we frequently noticed between chromosome ends in our preparations (Figs. 5 and 6) we suggest that the hybridization on the end of the B was due to rDNA which had been pulled off the X or the C in the process of squashing.

A large proportion of the micronucleoli in each preparation showed hybridization with rRNA (Figs. 5, 11, 12), but some were always free of silver grains. Even if all micronucleoli do contain rDNA and hybridize,

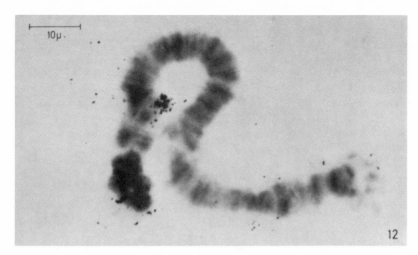

Fig. 12. Autoradiograph of the B chromosome from a *Rhynchosciara* salivary gland nucleus showing hybridization of radioactive *Xenopus* ribosomal RNA to a micronucleolus lying beside region B 15. Conditions of hybridization as described in Fig. 8. Exposure 89 days. ×1400

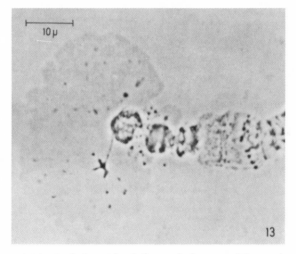

Fig. 13. Phase contrast photograph of the nucleolus organizing region of the X chromosome from a *Rhynchosciara* salivary gland nucleus. This chromosome bears a large nucleolus. The heterochromatic end of the X chromosome consists of 2 amorphous regions separated by a small beaded puff. ×1300

we should expect some to be unlabeled because of their small size and the few silver grains with which we are dealing. It is also possible that some of the small bodies do not contain rDNA. Labeled micronucleoli

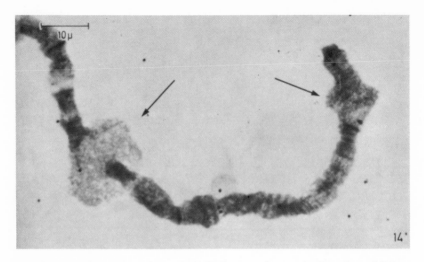

Fig. 14. Autoradiograph of chromosome II from a salivary gland nucleus of *Sciara coprophila* after hybridization with radioactive *Xenopus* ribosomal RNA. The two DNA puffs (arrows) show no hybridization to ribosomal RNA. The conditions of hybridization are those described in Fig. 8. Exposure 41 days. ×1300

were found along all of the chromosomes. They tended to clump preferentially around region 10 of chromosome A and region 15 of chromosome B (Figs. 5, 11, 12). These regions showed no grains when there were no micronucleoli.

We have seen no hybridization over any of the DNA puffs, nor any preferential association of micronucleoli with these regions in either *Rhynchosciara* or *Sciara* (Fig. 14). We conclude, therefore, that DNA puffs do not contain detectable amounts of rDNA.

3. Sciara coprophila

Our observations on *Sciara*, though not extensive, parallel the findings in *Rhynchosciara*. *Sciara coprophila* has a nucleolus near the centromere end of the X chromosome in younger larvae, but the nucleolus eventually disperses (Poulson and Metz, 1938; Crouse, 1960; Gabrusewycz-Garcia and Kleinfeld, 1966). Micronucleoli are scattered throughout the chromosome set (Gabrusewycz-Garcia and Kleinfeld, 1966). We found hybridization of *Xenopus* rRNA to the nucleolus organizer region of the X (Fig. 15) and to many of the micronucleoli. No other chromosome region was consistently labeled. In a few instances, however, the heavily labeled nucleolus organizer of the X was closely associated with the end of another chromosome. The *Sciara* slides which we have looked at thus far were

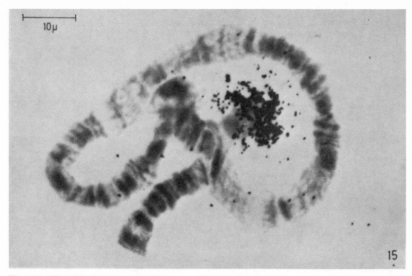

Fig. 15. The X chromosome from a salivary gland nucleus of *Sciara coprophila* showing hybridization of radioactive ribosomal RNA to the nucleolus organizing region at the end. The folding of this chromosome is due to the three repeat regions which it contains. The conditions of hybridization are described in Fig. 8. Exposure 45 days. ×1400

exposed .for a shorter time than the *Rhynchosciara* slides. Therefore we can not yet eliminate the possibility that *Sciara* has additional regions of rDNA.

4. Filter Hybridizations

DNA from adult *Rhynchosciara* was centrifuged to equilibrium on a CsCl gradient. The various gradient fractions were denatured with alkali and applied to individual nitrocellulose filters for hybridization according to the technique of Gillespie and Spiegelman (1965). *Xenopus* rRNA hybridized with DNA fractions that were slightly more dense than the main peak of the *Rhynchosciara* DNA (Fig. 16). We next carried out similar hybridization experiments in which the radioactive *Xenopus* rRNA was mixed with a large excess of unlabeled rRNA from either *Xenopus* or *Rhynchosciara*. In both cases the unlabeled rRNA effectively competed with the labeled *Xenopus* rRNA.

Discussion

Molecular hybridization experiments have shown that in *Drosophila melanogaster* and *Xenopus laevis* the number of cistrons coding for ribosomal RNA is proportional to the number of nucleolus organizers present in the genome. The original experiments of Ritossa and Spiegel-

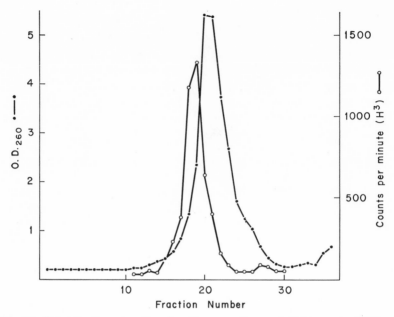

Fig. 16. Hybridization of radioactive *Xenopus* ribosomal RNA to the DNA of *Rhynchosciara*. DNA from adult *Rhynchosciara* was centrifuged to equilibrium in CsCl (•—•). Each gradient fraction was denatured and applied to a nitrocellulose filter. The filters were incubated with 2 μg/ml *Xenopus* ribosomal RNA H³ (5 × 10⁵ cpm/μg) in 6× SSC for 12 hours at 66° C. The filters were then treated with 20 μg/ml RNase in 2× SSC for 1 hour at 37° C. The ribosomal RNA (○—○) has hybridized to fractions on the denser side of the DNA peak

man (1965) indicated that the rDNA in *D. melanogaster* was located at or near the nucleolus organizer, since both the organizer and the ribosomal cistrons were located in the region between the break points of the sc⁴ and sc⁸ inversions. Later experiments (Ritossa, Atwood, Lindsley, and Spiegelman, 1965) showed that the ribosomal cistrons must lie at the *bobbed* locus in both the X and Y chromosomes. The *bobbed* locus has thus far proved inseparable from some nucleolus organizing activity (Cooper, 1959; Atwood, 1969). Although the location of the rDNA has not been so closely determined genetically in *Xenopus*, Wallace and Birnstiel (1966) and Birnstiel, Wallace, Sirlin, and Fischberg (1965) showed that loss of the visible nucleolus in the anucleolate mutant was accompanied by loss of the ribosomal cistrons. Furthermore, animals heterozygous for the loss of ribosomal cistrons have a nucleolus on only one member of the pair of nucleolar chromosomes. The simplest explanation for their observations is that the ribosomal cistrons are located at or near the nucleolus organizer in *Xenopus*, as in *Drosophila*.

Our experiments with *D. hydei*, *Rhynchosciara* and *Sciara* demonstrate that it is possible to use molecular hybridization to investigate the location of specific genes at the chromosome level. The technique permits the precision of cytological localization to be combined with the specificity of the hybridization procedure. In general our results confirm and extend the known relationship between the nucleolus organizer and the ribosomal cistrons. In addition they raise several questions about the organization of the rDNA in the chromosomes of Dipteran salivary gland cells.

The earlier studies of Kaufmann (1934, 1938) on *D. melanogaster* showed that the nucleolus organizer is located in the heterochromatic regions of the X and Y chromosomes. These regions become part of the chromocenter in salivary gland nuclei, and it has been reported that the nucleolus is connected to this chromocenter by a Feulgen-positive strand (Frolova, 1936; Emmens, 1937). In *D. hydei*, the species used in our study, Heitz (1934) showed that the nucleolus organizer is likewise located in the heterochromatin of the X and Y chromosomes and Spencer (1949) placed the *bobbed* locus at the end of the genetic map of the X. In our preparations hybridization of rRNA was detected only in the masses of DNA inside the nucleolus or occasionally in thin strands which could be seen to extend from the nucleolus. We have not seen hybridization within a distinct band on any chromosome. We conclude, therefore, that in salivary gland nuclei of *D. hydei* at late third instar the ribosomal cistrons are located almost exclusively within the body of the nucleolus. The physical relationship of the ribosomal cistrons to the rest of the DNA remains unclear.

In *Rhynchosciara* there is evident hybridization in the densely staining masses at one end of the X and the C chromosomes. In addition many of the micronucleoli show some hybridization. The sites of rDNA in the X and C chromosomes are densely staining areas with indefinite banding of the sort frequently referred to as heterochromatin. They often pair with each other and with similar areas in the A and B chromosomes (particularly A 11, B 15, and B 17). These regions are thought to be near the centromeres (Breuer and Pavan, 1955) although critical cytogenetic data on this point are apparently lacking. In *Sciara* we have thus far observed hybridization to the nucleolus organizer on the X and to the micronucleoli.

The occurrence of rDNA on the X of both sciarids was anticipated since a prominent nucleolus is found on the end of this chromosome in younger larvae (Breuer and Pavan, 1955; Gabrusewycz-Garcia and Kleinfeld, 1966; Mattingly and Parker, 1968). Hybridization on the C chromosome of *Rhynchosciara* was not expected, since a nucleolus had not been described on this chromosome. However it is not at all unusual for an organism to have two or more nucleolus organizers, and it may well be that a more careful look at the C chromosome at other stages or in other tissues will reveal some nucleolar activity (Beermann, 1960, 1962).

Hybridization in the micronucleoli calls for special consideration. Micronucleoli have been studied most extensively in *Sciara coprophila* (Swift, 1962; Gabrusewycz-Garcia and Kleinfeld, 1966). Gabrusewycz-Garcia and Kleinfeld have shown that micronucleoli may be found in close association with many bands along all the chromosomes. The micronucleoli contain RNA and they show incorporation of uridine; their fine structure is also similar to that of typical nucleoli. In lightly squashed salivary nuclei of *Sciara*, Gabrusewycz-Garcia (personal communication) has seen thin strands of DNA running among the polytene chromosomes and attached at places to micronucleoli. We also noted such strands in both *Sciara* and *Rhynchosciara*. Several of the strands which were in contact with the nucleolus organizer had hybridized with rRNA. Thus it is possible that in the intact nucleus the micronucleoli are part of a highly ramified nucleolus organizer. Our finding that the micronucleoli contain rDNA, but that the bands to which they are attached do not, is in keeping with this model.

Bodies somewhat similar to the micronucleoli of *Sciara* and *Rhynchosciara* have been detected in polytene nuclei of several flies, including *Sarcophaga bullata* (Whitten, 1965), *Chironomus melanotus* (Keyl and Hägele, 1966), *Hybosciara fragilis* (da Cunha, Pavan, Morgante, and Garrido, 1969), and *Dasyneura crataegi* (Henderson, 1967). The origin of these bodies has been ascribed to various regions of the chromosomes, including the nucleolus organizer and centromeres. On the basis of morphological evidence alone it is not possible to determine which of these bodies contain rDNA and which do not. On the other hand, such a determination should be relatively simple by cytological hybridization. It would seem useful to reserve the terms nucleolus and micronucleolus for those structures which contain rDNA.

In the Amphibian oocyte the production of multiple nucleoli involves the differential replication of the genes for ribosomal RNA (Brown and Dawid, 1968; Evans and Birnstiel, 1968; Gall, 1968, 1969). The same is true in the oocytes of several other organisms (reviewed in Gall, 1969). It has been suggested that the DNA in the micronucleoli of sciarid flies similarly reflects amplification of the ribosomal genes (da Cunha *et al.*, 1969). This suggestion is based on the presumed extrachromosomal location of the DNA and the observed incorporation of thymidine into the micronucleoli of *Hybosciara*. Although the hypothesis of gene amplification is an attractive one, the basic question is whether the ratio of rDNA to total DNA is greater in these cells than in cells which lack

74

micronucleoli. This question can only be resolved by quantitative hybridization experiments of the sort carried out in the oocyte systems.

Recently Ribbert and Bier (1969) have published a study of multiple nucleoli in the polytene nurse cells of *Calliphora erythrocephala*. These cells produce extrachromosomal nucleoli on a scale which appears to exceed that seen in the micronucleoli of sciarids. The nucleoli incorporate uridine and contain hundreds of Feulgen-positive bodies in the form of granules, fibers, or small rings. Because of the large amount of extra-chromosomal DNA and the likelihood that the nurse cells assume the role of ribosome formation for the oocyte (Bier, 1965; Bier, Kunz, and Ribbert, 1967), it seems probable that amplification of rDNA is present in these cells. Once again, however, quantitative hybridization studies are needed to settle the issue conclusively.

Mention should be made here of the well-known DNA puffs of *Sciara* and *Rhynchosciara*. As suggested originally by Breuer and Pavan (1955) and confirmed by the quantitative photometric studies of Rudkin and Corlette (1957), Swift (1962), and Crouse and Keyl (1968) these puffs undergo one or more replications beyond the level of polyteny character-istic of the chromosome as a whole. Our failure to find rDNA in the DNA puffs by the criterion of cytological hybridization is important in establishing that genes other than rDNA can undergo amplification. In all earlier studies where the amplified DNA was characterized by hybridization, rDNA was involved (reviewed in Gall, 1969).

Technical Aspects of the Cytological Hybridization Procedure

In our experiments with salivary gland nuclei we used radioactive RNA from tissue cultures of the toad, *Xenopus*, and complementary RNA copied from the rDNA satellite of *Xenopus* ovaries. We did this primarily because the cytological hybridization procedure requires RNA of higher specific radioactivity than we could obtain from Dipteran tissue. It is important, therefore, to consider the significance of the hybridization that occurs between species. Brown, Weber, and Sinclair (1967) studied the heterologous hybridization of rRNA in a large number of species. They found that *Xenopus* rRNA hybridized with the DNA of all eukaryotes investigated but failed to hybridize with the DNA of several bacteria and viruses. Experiments in this laboratory are con-sistent with these findings. We have, for instance, hybridized *Xenopus* rRNA with the DNA of several urodele Amphibians (unpublished), dytiscid beetles (Gall, Macgregor, and Kidston, 1969) and flies (Fig. 16). We have also hybridized mouse rRNA with *Xenopus* DNA (Gall, 1968). Neither mouse nor *Xenopus* rRNA hybridized with *E. coli* DNA. It is especially instructive to carry out the heterologous reactions with DNA fractionated by CsCl centrifugation. Such experiments show that the heterologous hybridization occurs in those gradient fractions that contain the rDNA (as determined by the homologous reaction) regardless of the base composition of the challenging RNA. A possible interpretation of this finding is that the rDNA of various eukaryotes contains conserved common sequences responsible for the cross reactions, and other non-conserved sequences which are responsible for the different base compositions (and buoyant densities). In any event it is possible to show that the heterologous reaction is competed by homologous rRNA.

As shown in Figure 16, *Xenopus* rRNA hybridizes with a fraction of *Rhynchosciara* DNA which is slightly more dense than the bulk of the *Rhynchosciara* DNA. In this case we have not carried out hybridization with *Rhynchosciara* rRNA to demonstrate that it, too, hybridizes with these fractions. However, unlabeled *Rhynchosciara* rRNA effectively competes with the heterologous reaction. It seems safe to conclude, therefore, that *Xenopus* rRNA hybridizes only with sequences contained in *Rhynchosciara* rDNA, even though there must be non-identical sequences in the rRNA of the two organisms. Insofar as the cytological hybridizations are concerned, it seems probable that any chromosome region in *Rhynchosciara* that hybridizes with *Xenopus* rRNA would also hybridize with *Rhynchosciara* rRNA. It is conceivable that the converse statement is not true, and that *Rhynchosciara* rRNA might hybridize with some additional chromosome regions. Further insight into this question could be obtained by hybridizing chromosomes with radioactive complementary RNA synthesized enzymatically using isolated *Rhynchosciara* rDNA as template.

We have made some changes in the hybridization procedure since our earlier reports (Gall and Pardue, 1969; Pardue and Gall, 1969). Experiments on *Xenopus* oocytes have shown that treating preparations with RNase before denaturation increased the level of hybridization without diminishing the specificity of the reaction. We have therefore used RNase pretreatment on our salivary gland preparations. The increase in hybridization brought about by RNase treatment appears to be greater in the salivary gland nuclei than in the *Xenopus* nuclei. Control salivary gland slides which had not been subjected to RNase digestion showed a specific but barely detectable labeling after four months of autoradiographic exposure. These results suggest that unlabeled RNA present in the cytological preparation may compete for binding sites in the DNA. In our original experiments we avoided RNase treatment before the hybridization step, because in filter hybridizations RNase binds to the filter and causes non-specific attachment of radioactive RNA (Gillespie and Spiegelman, 1965). It has recently been suggested by Das and Alfert (1969) that RNase might bind to cytological preparations and present similar difficulties in slide hybridization. Fortunately such non-specific binding apparently does not occur under our preparative conditions.

We have also found that the agar step which precedes denaturation (and which was used in most of the experiments reported here) is not essential. If the original tissue squash is very flat and thoroughly dried to the slide, then brief alkali treatment causes little morphological damage. If the nuclei are at all loose, however, they dissolve rapidly in alkali. The agar step merely eliminates the loss of less tightly bound material. We shall soon publish a detailed hybridization procedure with a discussion of technical features (Gall and Pardue, 1970).

It is not always possible to obtain RNA or DNA *in vivo* which is sufficiently radioactive for particular cytological hybridization experiments. In those cases where a specific DNA fraction can be isolated, enzymatic transcription of the DNA *in vitro* offers an alternative method of obtaining a radioactive hybridizing material. This alternative must, however, be used with caution since the RNA polymerase may transcribe certain sequences preferentially. We have used RNA transcribed from *Xenopus* ovarian satellite DNA (Reeder and Brown, 1969). Filter hybridization experiments show that satellite DNA from both ovary and somatic tissue binds about 20% of its weight of rRNA (Gall, 1969; Birnstiel, Speirs, Purdom, Jones, and Loening, 1968). Thus about 40% of the satellite is double stranded rDNA in the strict sense. The remaining 60% is a high $G + C$ DNA which is distributed among the ribosomal cistrons (Brown and Weber, 1968; Birnstiel *et al.*, 1968; Miller and Beatty, 1969). Since satellite DNA and cytologically detectable nucleolar DNA are probably synonymous (Gall, 1969), RNA copied from any part of either strand of the satellite should be satisfactory for the localization of regions containing rDNA.

We have characterized our complementary RNA by using it for slide hybridizations of *Xenopus* ovary preparations and by hybridizing it to fractions of a CsCl gradient of *Xenopus* DNA. In cytological preparations the complementary RNA hybridized with the nucleoli in early oogonia and with the mass of amplified rDNA in oocytes. Ribosomal RNA extracted from *Xenopus* tissue culture cells hybridized in the same places (Gall and Pardue, 1969; Pardue, 1969). In gradient hybridizations the complementary RNA hybridized primarily with the fractions containing the satellite DNA; however about 20% of the hybrid counts occurred in the main peak DNA. We have not yet determined the nature of the main peak hybridization.

In heterologous hybridizations with *Rhynchosciara* chromosomes, the complementary RNA labeled the same regions as did *Xenopus* rRNA, namely the ends of the X and C chromosomes, and the micronucleoli.

77

Thus from the standpoint of localizing the rDNA in cytological preparations, the complementary RNA appears to be sufficiently specific.

Recently two groups have reported success with cytological hybridization procedures somewhat different from ours. In both cases the DNA of the cytological preparation was denatured by heating. John *et al.* (1969) demonstrated hybridization of rRNA with the amplified rDNA of *Xenopus* oocytes. Their preparations of *Xenopus* oocytes were similar to those which we had hybridized after alkali denaturation (Gall, 1969; Gall and Pardue, 1969). They also hybridized HeLa tissue culture nuclei with total RNA derived from the same cell line. They briefly reported successful hybridization of polytene chromosomes using complementary RNA copied *in vitro* from total *Drosophila* DNA. Buongiorno-Nardelli and Amaldi (1969) developed a heat denaturation procedure for use on tissue sections. They were able to show hybridization of rRNA to the nucleolar region of interphase nuclei from the Chinese hamster. The intense labeling demonstrated by Buongiorno-Nardelli and Amaldi suggests that their technique may permit particularly efficient hybridization.

Summary

The organization of the cistrons coding for ribosomal RNA within the chromosomes of Dipteran salivary gland cells has been studied by cytological hybridization.

In *Drosophila hydei* the ribosomal cistrons are found within the body of the nucleolus. We have not been able to detect hybridization of ribosomal RNA with a distinct chromosomal band.

The DNA coding for ribosomal RNA in *Rhynchosciara hollaenderi* is localized on one end of the X chromosome and one end of the C chromosome. It is also found in many, possibly all, of the micronucleoli, which are scattered throughout the nucleus. An RNA has been prepared by *in vitro* transcription which hybridizes with these same regions of the salivary chromosomes in *Rhynchosciara*.

In *Sciara coprophila* ribosomal cistrons are found at the nucleolus organizer on the centromere end of the X chromosome, and in the micronucleoli.

The DNA which is amplified during the production of DNA puffs in the two sciarid flies apparently does not code for ribosomal RNA.

Acknowledgement. The technical assistance of Mrs. Cherry Barney is gratefully acknowledged.

References

Atwood, K. C.: Some aspects of the *bobbed* problem in *Drosophila*. Genetics **61** (Suppl.), 319—327 (1969).

Barr, H. J., Plaut, W.: Comparative morphology of nucleolar DNA in *Drosophila*. J. Cell Biol. **31**, C17—22 (1966).

Beermann, W.: Der Nukleolus als lebenswichtiger Bestandteil des Zellkerns. Chromosoma (Berl.) 11, 263—296 (1960).

— Riesenchromosomen. Protoplasmatologia, vol. VI/D. Wien: Springer 1962.

Bier, K.: Zur Funktion der Nährzellen im meroistischen Insektenovar unter besonderer Berücksichtigung der Oogenese adephager Coleopteren. Zool. Jb., Abt. allg. Zool. 71, 371—384 (1965).

— Kunz, W., Ribbert, D.: Struktur und Funktion der Oocytenchromosomen und Nukleolen sowie der Extra-DNS während der Oogenese panoistischer und meroistischer Insekten. Chromosoma (Berl.) 23, 214—254 (1967).

Birnstiel, M. L., Speirs, J., Purdom, I., Jones, K., Loening, U. E.: Properties and composition of the isolated ribosomal DNA satellite of Xenopus laevis. Nature (Lond.) 219, 454—463 (1968).

— Wallace, H., Sirlin, J. L., Fischberg, M.: Localization of the ribosomal DNA complements in the nucleolar organizer region of Xenopus laevis. Nat. Cancer Inst. Monogr. 23, 431—447 (1966).

Breuer, M. E., Pavan, C.: Behavior of polytene chromosomes of Rhynchosciara angelae at different stages of larval development. Chromosoma (Berl.) 7, 371—386 (1955).

Brown, D. D., Dawid, I. B.: Specific gene amplification in oocytes. Science 160, 272—280 (1968).

— Weber, C. S.: Gene linkage by RNA-DNA hybridization. II. Arrangement of the redundant gene sequences for 28 S and 18 S ribosomal RNA. J. molec. Biol. 34, 681—697 (1968).

— — Sinclair, J. H.: Ribosomal RNA and its genes during oogenesis and development. Carnegie Inst. Year Book 66, 580—589 (1967).

Buongiorno-Nardelli, M., Amaldi, F.: Autoradiographic detection of molecular hybrids between H³-rRNA and DNA present in tissue sections. (In press.)

Burgess, R. R., Travers, A. A., Dunn, J. J., Bautz, E. K. F.: Factor stimulating transcription by RNA polymerase. Nature (Lond.) 221, 43—46 (1969).

Conger, A. D., Fairchild, L. M.: A quick-freeze method for making smear slides permanent. Stain Technol. 28, 281—283 (1953).

Cooper, K. W.: Cytogenetic analysis of major heterochromatic elements (especially Xh and Y) in Drosophila melanogaster and the theory of "heterochromatin". Chromosoma (Berl.) 10, 535—588 (1959).

Crouse, H. V.: The controlling element in sex chromosome behavior in Sciara. Genetics 45, 1429—1443 (1960).

— Keyl, H.-G.: Extra replications in the "DNA puffs" of Sciara coprophila. Chromosoma (Berl.) 25, 357—364 (1968).

Cunha, A. B. da, Pavan, C., Morgante, J. S., Garrido, M. C.: Studies on cytology and differentiation in Sciaridae. II. DNA redundancy in salivary gland cells of Hybosciara fragilis (Diptera, Sciaridae). Genetics 61 (Suppl.), 335—349 (1969).

Das, N. K., Alfert, M.: Binding of labeled ribonucleic acid to basic proteins, a major difficulty in ribonucleic acid-deoxyribonucleic acid hybridization in fixed cells in situ. J. Histochem. Cytochem. 17, 418—422 (1969).

Emmens, C. W.: The morphology of the nucleus in the salivary glands of four species of Drosophila. Z. Zellforsch. 26, 1—20 (1937).

Evans, D., Birnstiel, M.: Localization of amplified ribosomal DNA in the oocyte of Xenopus laevis. Biochim. biophys. Acta (Amst.) 166, 274—276 (1968).

Frolova, S. L.: Structure of the nuclei in the salivary gland cells of Drosophila. Nature (Lond.) 137, 319 (1936).

Gabrusewycz-Garcia, N., Kleinfeld, R. G.: A study of the nucleolar material in *Sciara coprophila*. J. Cell Biol. **29**, 347—359 (1966).

Gall, J. G.: Differential synthesis of the genes for ribosomal RNA during amphibian oogenesis. Proc. nat. Acad. Sci. (Wash.) **60**, 553—560 (1968).

— The genes for ribosomal RNA during oogenesis. Genetics **61** (Suppl.), 121—132 (1969).

— Macgregor, H. C., Kidston, M. E.: Gene amplification in the oocytes of Dytiscid water beetles. Chromosoma (Berl.) **26**, 169—187 (1969).

— Pardue, M. L.: Formation and detection of RNA-DNA hybrid molecules in cytological preparations. Proc. nat. Acad. Sci. (Wash.) **63**, 378—383 (1969).

— — Nucleic acid hybridization in cytological preparations. In: Methods in enzymology XII C (L. Grossman and K. Moldave, eds.). New York: Academic Press Inc. (in preparation).

Gillespie, D., Spiegelman, S.: A quantitative assay for DNA-RNA hybrids with DNA immobilized on a membrane. J. molec. Biol. **12**, 829—842 (1965).

Heitz, E.: Die somatische Heteropyknose bei *Drosophila melanogaster* und ihre genetische Bedeutung. Z. Zellforsch. **20**, 237—287 (1934).

— Bauer, H.: Beweise für die Chromosomennatur der Kernschleifen in den Knäuelkernen von *Bibio hortulanus* L. Z. Zellforsch. **17**, 67—82 (1933).

Henderson, S. A.: The salivary gland chromosomes of *Dasyneura crataegi* (*Diptera*: *Cecidomyiidae*). Chromosoma (Berl.) **23**, 38—58 (1967).

Jacob, J., Sirlin, J. L.: Electron microscope studies on salivary gland cells. I. The nucleus of *Bradysia mycorum* Frey (*Sciaridae*), with special reference to the nucleolus. J. Cell Biol. **17**, 153—165 (1963).

John, H. A., Birnstiel, M. L., Jones, K. W.: RNA-DNA hybrids at a cytological level. Nature (Lond.) **223**, 582—587 (1969).

Kaufmann, B. P.: Somatic mitoses of *Drosophila melanogaster*. J. Morph. **56**, 125—155 (1934).

— Nucleolus-organizing regions in salivary gland chromosomes of *Drosophila melanogaster*. Z. Zellforsch. **28**, 1—11 (1938).

Keyl, H.-G., Hägele, K.: Heterochromatin-Proliferation an den Speicheldrüsen-Chromosomen von *Chironomus melanotus*. Chromosoma (Berl.) **19**, 223—230 (1966).

Mattingly, E., Parker, C.: Sequence of puff formation in *Rhynchosciara* polytene chromosomes. Chromosoma (Berl.) **23**, 255—270 (1968).

Miller, O. L., Jr., Beatty, B. R.: Visualization of nucleolar genes. Science **164**, 955—957 (1969).

Olvera, R. O.: The nucleolar DNA of three species of *Drosophila* in the hydei complex. Genetics **61** (Suppl.), 245—249 (1969).

Pardue, M. L.: Nucleic acid hybridization in cytological preparations. J. Cell Biol. **43**, 101a (1969).

— Gall, J. G.: Molecular hybridization of radioactive DNA to the DNA of cytological preparations. Proc. nat. Acad. Sci. (Wash.) **64**, 600—604 (1969).

Pavan, C.: Nucleic acid metabolism in polytene chromosomes and the problem of differentiation. Brookhaven Symp. in Biol. **18**, 222—241 (1965).

Poulson, D. F., Metz, C. W.: Studies on the structure of nucleolus-forming regions and related structures in the giant salivary gland chromosomes of Diptera. J. Morph. **63**, 363—395 (1938).

Reeder, R. H., Brown, D. D.: The fidelity of transcription of *Xenopus* ribosomal RNA genes by bacterial RNA polymerase. J. Cell Biol. **43**, 114a—115a (1969).

Ribbert, D., Bier, K.: Multiple nucleoli and enhanced nucleolar activity in the nurse cells of the insect ovary. Chromosoma (Berl.) **27**, 178—197 (1969).

Ritossa, F. M., Atwood, K. C., Lindsley, D. L., Spiegelman, S.: On the chromosomal distribution of DNA complementary to ribosomal and soluble RNA. Nat. Cancer Inst. Monogr. **23**, 449—472 (1966).

— Spiegelman, S.: Localization of DNA complementary to ribosomal RNA in the nucleolus organizer region of *Drosophila melanogaster*. Proc. nat. Acad. Sci. (Wash.) **53**, 737—745 (1965).

Rodman, T. C.: Morphology and replication of intranucleolar DNA in polytene nuclei. J. Cell Biol. **42**, 575—582 (1969).

Rudkin, G. T., Corlette, S. L.: Disproportionate synthesis of DNA in a polytene chromosome region. Proc. nat. Acad. Sci. (Wash.) **43**, 964—968 (1957).

Spencer, W. P.: Gene homologies and the mutants of *Drosophila hydei*. In: Genetics, paleontology, and evolution (G. L. Jepsen, E. Mayr, and G. G. Simpson, eds.). Princeton: Princeton University Press (1949).

Swift, H.: Nucleic acids and cell morphology in Dipteran salivary glands. In: The molecular control of cellular activity (J. M. Allen, ed.). New York: McGraw-Hill Book Co., Inc. 1962.

Vincent, W. S., Miller, O. L., Jr. (eds.): International symposium on the nucleolus-its structure and function. Nat. Cancer Inst. Monogr. **23** (1966).

Wallace, H., Birnstiel, M. L.: Ribosomal cistrons and the nucleolar organizer. Biochim. biophys. Acta (Amst.) **114**, 296—310 (1966).

Whitten, J. M.: Differential deoxyribonucleic acid replication in the giant foot-pad cells of *Sarcophaga bullata*. Nature (Lond.) **208**, 1019—1021 (1965).

The 5 s RNA Genes of *Drosophila melanogaster*

KENNETH D. TARTOF AND ROBERT P. PERRY

1. Introduction

The ribosome, from whatever source thus far examined, contains one molecule of 5 s RNA as well as one molecule each of 16 to 18 s and 23 to 28 s RNA (cf. Aubert, Monier, Reynier & Scott, 1967). In *Drosophila*, there are approximately 130 or more copies of each 18 and 28 s RNA gene per haploid genome, and these are clustered at the nucleolus organizer loci on the sex chromosomes (Ritossa, Atwood & Spiegelman, 1966b). Because of the functional relationships among the various ribosomal RNA's and the apparent need to co-ordinate their syntheses, it seemed of considerable interest to determine the number of 5 s rRNA† genes in *Drosophila*, and to define their linkage relationship to the other rRNA genes.

In *Bacillus subtilis* evidence has been presented indicating that the 5, 16 and 23 s rRNA genes are tightly clustered, and that there are approximately 4, 10 and 10 copies of each gene per genome, respectively (Smith, Dubnau, Morell & Marmur, 1968). In a similar study of rRNA genes Brown & Weber (1968) concluded that in the toad, *Xenopus laevis*, there are more than 60 times as many 5 s genes as 18 and 28 s genes, and that the 5 s genes are not intermingled with those specifying the 18 and

† Abbreviation used: rRNA, ribosomal RNA.

28 s components. Our results indicate that in *Drosophila* the number of genes coding for 5 s rRNA is roughly the same as for the other rRNA species. Moreover, it appears that the 5 s genes are not located on the sex chromosomes, and hence that they are not linked to the other rRNA genes.

2. Materials and Methods

(a) *Drosophila stocks*

Drosophila melanogaster were raised at $24\cdot0 \pm 0\cdot5°C$ in half-pint bottles using the standard corn meal–agar medium. An Oregon-R wild-type strain was supplied by Dr Jack Schultz. In experiments designed to determine whether the 5 s RNA genes are localized in the nucleolus organizer region we utilized $In(1)\ sc^{4L}sc^{8R}/Y$ males possessing one Y chromosome and an X with a deletion for the proximal one-third of its length, which includes the nucleolus organizer locus. To determine if the 5 s RNA genes are located in any other portion of the X we used C(1) DX/Y females. Such females possess a Y and a compound double X which consists of two X chromosomes tandemly linked together. This compound X is duplicated for all the genes of the X chromosome except the following: (1) the deleted region, referred to above; (2) one of the minute heterochromatic arms to the right of the centromere; (3) an extremely small heterochromatic region immediately to the left of the centromere; (4) a very small region ($<0\cdot01$ map unit) at the extreme left tip and (5) possibly a small portion of the heterochromatic region, designated hD (Cooper, 1959).

Males and females of the above genotypes are both present in the stock designated $In(1)sc^{4L}sc^{8R},\ y\ sc^4sc^8\ cv\ v\ B/C(1)\ DX,\ yf/B^SY$, obtained from the collection at the Oak Ridge National Laboratory. A complete description of these mutants has been compiled by Lindsley & Grell (1968). For the sake of brevity and clarity in this report $In(1)\ sc^{4L}sc^{8R}/Y$ males are symbolized as $sc^4\text{-}sc^8/Y$ ♂ and C(1) DX/Y females as $\overline{\text{XX}}/\text{Y}$ ♀. The $sc^4\text{-}sc^8/Y$ ♂ and $\overline{\text{XX}}/\text{Y}$ ♀, respectively, contain about 97 and 104% of the DNA in the wild-type genome. It should be noted that although $sc^4\text{-}sc^8/Y$ males generally tend to accumulate an extra Y chromosome due to non-disjunction, this condition does not occur with the genetic system used here because both the X and Y chromosomes carry the bar eye gene (B), which is lethal when present in a triple dose.

(b) *Radioactive labeling*

Ore-R eggs, collected over a 5-hr laying period, were placed on 15 ml. of a low yeast medium (Ritossa, Atwood & Spiegelman, 1966a) to which 5 mc-[5-^{3}H]uridine (New England Nuclear Corp., 28·3 c/m-mole) was added. After 9 days growth, larvae were collected by adding a solution of saturated NaCl to each bottle, stirring and pooling the contents of all bottles in a large beaker. The larvae, which float in NaCl of this density, are then scooped off the surface with a spoon and washed once with water.

(c) *Preparation of 4 and 5 s RNA*

10 g of larvae were homogenized at 0°C in a Teflon Potter–Elvehjem homogenizer in 15 ml. of Tris-NaCl buffer (0·05 M-Tris, pH 7·0, 0·2 M-NaCl) in 1·0% sodium dodecyl sulfate and in 25 ml. of Tris–NaCl buffer-saturated phenol containing 0·1% 8-hydroxy-quinoline. The emulsion was broken by centrifugation for 10 min at 8000 g.

The aqueous phase and interphase were mixed with equal volumes of phenol and chloroform–isoamylalcohol (24:1, v/v), and then shaken at room temperature for 20 min. Following centrifugation the aqueous phase was repeatedly extracted with phenol at room temperature until no interphase remained. After removal of the phenol with ether, and flushing with nitrogen gas, solid NaCl was added to a final concentration of 1·5 M, and the solution allowed to stand overnight at 2°C. After centrifugation, the precipitate, which contains predominantly 18 and 28 s RNA, was dissolved in a small amount of Ac–NaCl–EDTA buffer (0·01 M-acetate, pH 6·0, 0·1 M-NaCl, 0·001 M-EDTA) and stored at −30°C to await further purification as described below. The supernatant, containing the 4 s and 5 s RNA, was dialyzed at 2°C against 10 vol. of Ac–NaCl–EDTA buffer, changed 3 times every 4 hr.

83

The dialyzate was centrifuged for 30 min at 105,000 g to remove glycogen. RNA was precipitated by addition of 2 vol. of ethanol and allowed to stand overnight at $-30°C$. The precipitated RNA, collected by centrifugation, was then dissolved in Ac–NaCl–EDTA buffer containing 20% sucrose to a final concentration of 10 mg/ml. For preparative separation of 4 and 5 s RNA 50 μl. portions of this mixture were layered on top of 6·8% polyacrylamide gels and submitted to electrophoresis under the conditions described below. The gels were then removed from their plastic moulds and examined under oblique ultraviolet illumination (short-wave setting, Mineralight, UVSL-25, Ultra-Violet Products, Inc.). The segments containing the 4 s RNA and those containing 5 s RNA (readily visualized as two fast migrating bands separated by about 1·5 cm) were cut out with a razor blade, frozen on dry ice, and cut into 1-mm sections with a manifold of razor blades. Slices from the gel segments containing 4 or 5 s RNA were placed in Ac–NaCl–EDTA buffer containing 0·5% sarkosyl NL97 (Geigy Laboratory, recrystallized once from ethanol) and shaken continuously at 2°C. This solution was changed 3 times every 16 hr, and the pooled extracts were centrifuged at 14,000 g for 10 min. The supernatant solution was passed through a nitrocellulose filter (Schleicher & Schuell, B6, 24 mm, previously soaked in Ac–NaCl–EDTA buffer) added to 2 vol. of ethanol, and allowed to stand at $-30°C$ for 48 hr. The precipitated RNA was dissolved in $2 \times$ SSC (SSC is 0·15 M-NaCl, 0·15 M-sodium citrate, pH 7·2) and stored at $-30°C$. The RNA was assayed by measuring the absorbance at 260 nm, using as a zero absorbance reference, sarkosyl extracts of a blank gel treated in an identical manner. The yield from such extractions was usually about 45%.

(d) *Preparation of 18 and 28 s RNA*

The redissolved 1·5 M-Nacl precipitate (60 A_{260} units) containing $18+28$ s RNA (see above) were layered onto 15 to 30% (w/v) sucrose gradients prepared in Ac–NaCl–EDTA buffer. After centrifuging for 19 hr at 27,000 rev./min in a Spinco SW27 rotor at 3°C, the RNA was passed through the flow cell of a Gilford recording spectrophotometer and fractionated in the cold into 1-ml. portions. Pooled fractions containing the 18 or 28 s RNA components were reprecipitated with ethanol and submitted to a second density gradient centrifugation. Samples of labeled 18 or 28 s RNA or unlabeled $18 + 28$ s RNA were precipitated with ethanol and redissolved in small volumes of $2 \times$ SSC.

(e) *Preparation of L-cell RNA*

[32]P-labeled RNA's were prepared from L-cell ribosomes by methods described previously (Perry & Kelley, 1968).

(f) *Polyacrylamide gels*

Analysis of RNA on acrylamide gels was performed according to Weinberg, Loening, Willems & Penman (1967) except that no glycerol was used. 4 and 5 s RNA was separated on 6·8% acrylamide–0·28% diacrylate gels, and 18, 28 and 38 s RNA on 2·7% acrylamide–0·25% diacrylate gels in E buffer (0·04 M-Tris, 0·02 M-sodium acetate, 0·001 M-EDTA, pH 7·2). Electrophoresis was performed at 5 mA/gel in E buffer containing 2% sodium dodecyl sulfate. Preparative electrophoresis of 4 and 5 s RNA was performed for 9 hr on gels 9·2-cm long, 0·9-cm diameter.

Gels were scanned at 260 nm with the use of a Gilford Linear Transport gel scanning attachment. [32]P radioactivity was measured by slicing frozen gels into 1-mm sections and drying them on glass fiber filters (Reeve Angle, 934AH, 2·4 cm) at 60°C. The filters were then placed in 5 ml. of fluid containing Liquiflor (New England Nuclear Corp.) diluted 1:25 with toluene, and counted in a scintillation counter. Gel sections containing ^{3}H label were dissolved in NH_4OH (Penman, Vesco & Penman, 1968), then mixed with 10 ml. of scintillation fluid (Perry & Kelley, 1968) and counted.

(g) *Preparation of DNA*

Adult flies of the appropriate genotype were lightly etherized and their DNA immediately extracted according to the methods of Ritossa, Atwood & Spiegelman (1966b).

(h) *DNA–RNA hybridization*

The DNA was denatured and adsorbed to nitrocellulose filters as described by Ritossa, Atwood & Spiegelman (1966*b*).

For hybridization reactions, each filter was incubated with a given concentration of RNA in a glass vial containing 3·0 ml. 2 × SSC at 60±0·1°C. More than one filter per vial gave erratic results. However, a vial containing the same RNA solution could be used repeatedly with fresh DNA-loaded nitrocellulose filters. As many as 8 incubations of a single RNA sample did not result in any detectable difference in the amount of RNA hybridized. The incubations with ³H-labeled 18 and 28 s RNA were carried out for 12 hr, and incubations with ³H-labeled 4 or 5 s RNA, for 2 hr. These incubation periods are sufficient to give maximum hybridization of the RNA species employed (Ritossa & Spiegelman, 1965; Ritossa, Atwood, Lindsley & Spiegelman, 1966). In preliminary experiments with 5 s RNA we verified that the hybridization reaction was complete after 2 hr and remains unchanged for at least 12 hr.

The incubated filters were rinsed by suction filtration with 50 ml. 2 × SSC on each side, placed in 5·0 ml. of 2 × SSC containing 20 μg RNase/ml. (5 times crystallized, Sigma Biochemicals), and allowed to stand at room temperature for 60 min. Each filter was again rinsed on both sides with 50 ml. 2 × SSC, dried and counted in a Beckman LS250 scintillation counter. A background subtraction was made using filters treated in an identical manner but containing no DNA. The background values were ≦20 cts/min even when the filters were incubated with more than 10⁶ cts/min.

(i) *Nucleotide composition*

To gel slices containing selected RNA components 500 μg yeast transfer RNA and 1·0 ml. of 3 N-KOH were added to a final volume of 10 ml. After incubation for 18 hr at 37°C, 5 ml. of the hydrolysate was loaded on a Dowex-1 (X8) formate column and the nucleotides were separated as previously described (Petrović & Janković, 1962). Nucleotides were dried on planchets and counted in a gas-flow counter.

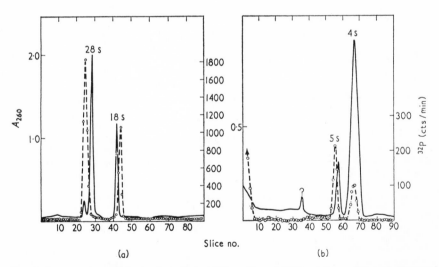

FIG. 1. A comparison of electrophoretic mobilities of larval RNA of *Drosophila* and ribosomal RNA of L cells.

(a) *Drosophila* 28 + 18 s RNA (———; A_{260}) from the 1·5 M-NaCl precipitate (see Materials and Methods) was co-run with ³²P-labeled RNA extracted from L-cell ribosomes (--O--O--) on 2·7% polyacrylamide gels for 4 hr at 5 mA/gel.

(b) *Drosophila* RNA (———; A_{260}) from the dialyzed 1·5 M-NaCl supernatant co-run with the same preparation of L-cell rRNA used in (a) (-- O -- O --) on 6·8% polyacrylamide gels.

3. Results

(a) *Identification and properties of various* Drosophila *RNA's*

Figure 1 shows the results of polyacrylamide gel electrophoresis of *Drosophila* larval RNA co-run with ^{32}P-labeled RNA from L-cell ribosomes. In agreement with an earlier report by Loening (1968), the higher molecular weight rRNA's of *Drosophila* differ in their electrophoretic mobilities from those of mammalian-rRNA (Fig. 1(a)). In Figure 1(b) it is seen that the 5 s component of *Drosophila* also has a slightly different mobility than that of L-cell 5 s rRNA, although the 4 s components from these two animal species migrate coincidentally. The minor component (?) migrating more slowly than 5 s RNA has been regularly observed; however its significance is at present unknown.

The nucleotide compositions of ^{32}P-labeled 28, 18, 5 and 4 s RNA purified by polyacrylamide gel electrophoresis are compared in Table 1. The base compositions of *Drosophila* 28 and 18 s RNA reported here are in close agreement with similar analyses by other workers (Hastings & Kirby, 1966; Ritossa, Atwood, Lindsley & Spiegelman, 1966; Schultz & Travaglini, personal communication) and thus serve as a check on the method used in these studies. It is seen that both the 4 and 5 s RNA's

TABLE 1

Nucleotide compositions of various RNA's of Drosophila melanogaster
(values presented as means $\pm S.E.$)

RNA	N	Cytosine	Adenine	Uracil	Guanine	% G + C
28 s	5	15·6±0·6	28·1±1·1	33·8±1·8	22·5±0·6	38·1±1·0
18 s	5	16·5±0·7	26·5±0·9	33·2±2·2	24·4±0·8	40·9±1·4
4 s	5	26·6±0·8	18·3±0·4	22·5±2·0	32·4±1·2	59·0±1·9
5 s	5	23·9±1·4	20·9±0·4	26·7±1·5	28·5±0·3	52·4±1·6
38 s	2	15·1±0·7	29·3±0·7	38·0±1·4	17·6±0·5	32·7±0·7
38 s†		13·7	30·0	40·0	17·3	31·0

For 4, 5, 18 and 28 s RNA, 48-hr old Ore-R larvae were placed for 24 hr on 15 ml. of modified medium containing 1·5 mc-[^{32}P]phosphoric acid (carrier free, International Chemical and Nuclear Corp.). For the 38 s RNA component, 50 adult Ore-R females were injected with 45 μc (0·1 μl.) [^{32}P]phosphoric acid/fly and harvested 3 hr later. RNA's were extracted as described in the Materials and Methods section except that NaCl precipitation was omitted. After ethanol precipitation the adult RNA was directly submitted to electrophoresis on 2·7% polyacrylamide gels; larval RNA was first fractionated by density gradient centrifugation, after which the 4 to 5 s components (2 to 9 s region of the gradient) and the 18 to 28 s components were submitted to electrophoresis on 6·8 and 2·7% polyacrylamide gels, respectively. The gels were sliced and monitored in a gas-flow counter. Fractions corresponding to the homogeneous peaks of 4, 5, 18, 28 and 38 s RNA were hydrolyzed as described in the Materials and Methods section.

N is the number of determinations.

† Corrected for 22% contamination with a heterogeneous pulse-labeled species. The amount and base composition of the contaminating species were respectively estimated from the acrylamide gel profile and a nucleotide analysis of the radioactive RNA on the high molecular weight side of the 38 s precursor peak.

From hybridization data (presented later) one can estimate that the maximum amount of contamination of the 4 and 5 s RNA samples in the fragments of 18 and 28 s rRNA is less than 10%. Since corrections of the base compositions of the 4 and 5 s RNA for this amount of contamination are less than the standard errors of the measurements, no correction was made.

are composed of a considerably higher proportion of guanylic plus cytidylic residues than either the 28 or 18 s rRNA of *Drosophila*. For comparative purposes we have also measured the base composition of the 38 s precursor of the rRNA components (Edström & Daneholt, 1967; Greenberg, 1969), and find that it is even more (A+U)-rich than its 28 and 18 s RNA products.

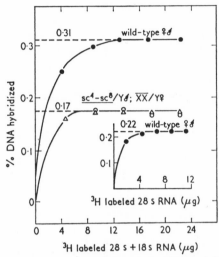

FIG. 2. Saturation of *Drosophila* DNA with 28 + 18 s and 28 s ribosomal RNA.

Filters containing 75 µg adult DNA isolated from a mixture of wild-type males and females (●), or $\overline{XX}$/Y females (△), or 27 µg of adult DNA from sc^4-sc^8/Y males (○) were incubated for 12 hr at 60°C in 3 ml. of 2 × SSC containing the indicated amounts of 28 + 18 s ³H-labeled RNA (62,700 cts/min/µg).

The inset shows the results of incubating filters containing 75 µg of wild-type DNA with purified ³H-labeled 28 s RNA under identical conditions. The saturation plateaux for wild-type DNA were approximately 11,000 and 15,000 cts/min/filter with 28 s and 28 + 18 s RNA, respectively. For $\overline{XX}$/Y and sc^4-sc^8/Y stocks the plateaux correspond to 6500 and 2800 cts/min/filter, respectively. Background was 20 cts/min/filter.

TABLE 2

Estimate of the number of 4, 5, 28, and 28+18 s RNA genes per haploid genome in Drosophila melanogaster

% DNA hybridized	4 s	5 s	28 s	28 s + 18 s
	0·015	0·0065–0·0077	0·22	0·31
DNA hybridized (daltons)†	$1·8 \times 10^7$	$7·80$–$9·24 \times 10^6$	$2·64 \times 10^8$	$3·72 \times 10^8$
Molecular weight of RNA (daltons)	$2·4 \times 10^4$	$4·0 \times 10^4$	$1·4 \times 10^6$	$2·1 \times 10^6$
Number of genes	750	195–230	190	180

† Assuming the molecular weight of the haploid genome in *Drosophila melanogaster* to be $1·2 \times 10^{11}$ daltons (Rudkin, personal communication and *Proc. 11th Int. Genetics Congr.* p. 359, 1966).

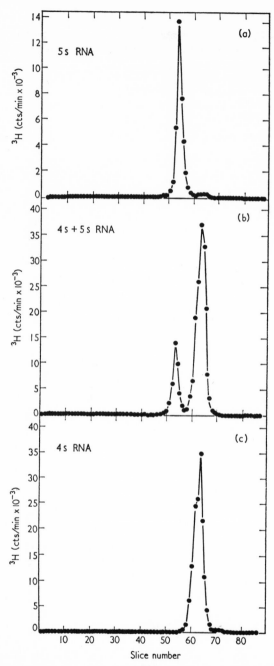

Fig. 3. Electrophoresis of a sample of purified 5 s RNA (a), 4 s RNA (c) and a mixture of each (b). Purification was achieved according to the details given in Materials and Methods, and the purity confirmed by a second electrophoresis on 6·8% polyacrylamide gels for 3 hr at 5 mA/gel.

(b) *Hybridization of 28 and 18 s ribosomal RNA*

As an initial check on our hybridization procedures we measured the proportion of DNA complementary to 28 s or 28+18 s rRNA (Fig. 2). With DNA from wild-type flies saturation plateaux equivalent to 0·22 and 0·31% of the DNA were obtained for 28 s and 28+18 s RNA, respectively. These values are in good agreement with those previously published (Ritossa & Spiegelman, 1965), and yield comparable estimates of the number of 28 and 18 s genes (Table 2). Moreover, 28+18 s RNA hybridizes to 0·17% of the DNA from $\overline{XX}$/Y females or sc^4-sc^8/Y males, thus confirming the genetic analysis which indicates that these flies contain only one nucleolar organizer, viz. that on their Y chromosome.

(c) *Purity of low molecular weight* Drosophila *RNA's*

When the preparations of purified ^{3}H-labeled *Drosophila* 4 and 5 s RNA used in the hybridization reactions were submitted to a second electrophoresis on polyacrylamide gels, they were found to migrate as single bands with a maximum of 2% cross contamination (Fig. 3). Nevertheless, initial hybridization experiments utilizing relatively high concentrations of purified 5 s RNA (up to 6 µg) failed to yield a saturation plateau. A similar situation was encountered by Ritossa, Atwood & Spiegelman (1966a) in their attempts to determine the number of tRNA genes in *Drosophila*. They attributed such "noise" to contaminating fragments of 28 and 18 s RNA,† and were able to obtain approximate plateaux when hybridizations were carried out in the presence of excess amounts of unlabeled 28+18 s RNA. The same strategy was used in the present study. As shown in Fig. 4, the addition of a large excess of unlabeled *Drosophila* 28+18 s RNA reduces the apparent level of hybridization by more than

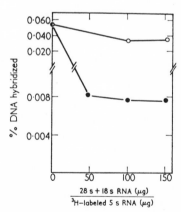

FIG. 4. Competition of ^{3}H-labeled 5 s RNA with unlabeled 28+18 s RNA from either L cells (—○—○—) or *Drosophila* (—●—●—).

Each point represents a filter containing approximately 75 µg of adult Ore-R (♀♂) *Drosophila* DNA which was incubated for 2 hr at 60°C in 3 ml. of 2 × SSC containing 3·6 µg of ^{3}H-labeled 5 s RNA (55,000 cts/min/µg) and the indicated amount of unlabeled 28 + 18 s RNA.

† Such fragments would continue to migrate with the 4 and 5 s components, and hence would not be detected on the electrophoresis rerun.

sixfold. In contrast the addition of equivalent quantities of L-cell ribosomal RNA reduces the hybridization level by only about 30%, indicating that for the most part the unlabeled *Drosophila* RNA is competitively replacing a ^{3}H-labeled 28+18 s RNA contaminant. In all subsequent hybridization reactions a 100 to 1 ratio of unlabeled *Drosophila* 28+18 s RNA to ^{3}H-labeled 5 s RNA was used.

(d) *Distribution and number of genes for 5 s RNA*

Under the conditions described above, plateaux with ^{3}H-labeled 5 s RNA over a twofold concentration range were achieved (Fig. 5(a)). For a mixed population of wild-type males and females, a saturation value of 0·0077% of the DNA hybridized to 5 s RNA was found. A *t*-test indicated that there was no significant difference between the plateau value of males and females.

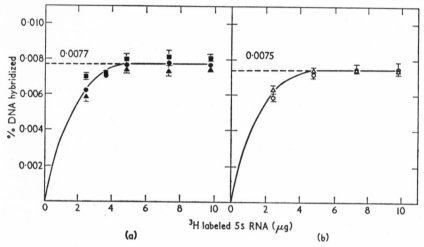

FIG. 5. Saturation of *Drosophila* DNA with ^{3}H-labeled 5 s RNA.
Filters containing 75 μg of adult DNA extracted from (a) wild-type males (▲), females (■), or a mixture of both (●) or (b) $\overline{XX}$/Y females (○) or *sc⁴-sc⁸*/Y males (△) were incubated for 2 hr at 60°C in 3 ml. of 2×SSC containing the appropriate concentration ^{3}H-labeled 5 s RNA (55,000 cts/min/μg) and 100 times that amount of cold 28 + 18 s RNA. Each point represents the mean ± standard error of two independent determinations. At the saturation plateau the average filter contained approximately 320 cts/min with a background (i.e. similarly treated filter containing no DNA) of 20 cts/min.

Since males and females have an equal number of 5 s RNA genes, one may conclude that if such genes were located on the sex chromosomes then they must be evenly distributed between the X and Y. This possibility is tested in the hybridization experiments shown in Figure 5(b). Males carrying the *sc⁴-sc⁸* deletion in their X chromosome have saturation values which are essentially identical to wild-type males and females, thus indicating that the 5 s genes are not located in the deleted region. If the 5 s genes were located on the remaining portion of the sex chromosomes, then the DNA from $\overline{XX}$/Y females should exhibit a higher saturation plateau than the other genotypes tested. Since $\overline{XX}$/Y flies are practically duplicated for the remainder of the X

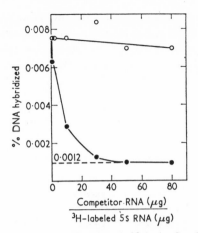

Fig. 6. Competition of ^{3}H-labeled 5 s RNA with cold 5 s (—●—●—) or 4s (—○—○—) RNA. Filters containing 75 μg of DNA extracted from Ore-R (♀♂) were incubated for 2 hr at 60°C in 3 ml. of 2 × SSC containing 3·6 μg ^{3}H-labeled 5 s RNA (55,000 cts/min/μg), 360 μg of cold 28 + 18 s RNA and the indicated amount of unlabeled competitor 4 or 5 s RNA. Saturation levels of 0·0077 and 0·0012% of the DNA hybridized correspond to approximately 320 cts/min/filter and 45 cts/min/filter, respectively. Background was 20 cts/min/filter.

chromosome,† they would be expected to give a 50% higher saturation value if the 5 s genes resided in these duplicated segments, a difference which would readily be detected in our experiments. However, no difference was observed. This analysis leads to the conclusion that the 5 s RNA genes of *Drosophila melanogaster* are not located exclusively on the sex chromosomes. If the 5 s RNA genes were evenly distributed on all the chromosomes then about 20% would reside on the sex chromosomes, which account for roughly 20% of the DNA. Hence it may be concluded that at least 80% of the genes for 5 s RNA are located in the autosomes.‡ If the 5 s RNA genes are grouped in a single cluster, then our results indicate that this cluster must be on one of the autosomes.

On the premise that the sites on the DNA-filters which bind the 5 s RNA are truly specific for this RNA species, one may use the saturation plateaux of Figure 5 to determine the number of 5 s RNA genes in *Drosophila*. The flatness of the plateau indicates that, under the hybridization conditions used here, there is little or no binding to non-specific sites on the DNA-filters, for such noise would be associated with a failure to achieve flat plateaux. To substantiate further this point, and to obtain a

† As mentioned earlier, the $\overline{\text{XX}}$/Y chromosome contains only one dose of a few small heterochromatic regions and one dose of the extreme left tip of the X. If the 5 s genes were located in these regions, then no differences in the saturation plateau would be detected with the genotypes used. However, location of the 5 s genes in two of these regions is considered unlikely for the following reasons. (1) The genes located in the left tip of the X have no homologous region in the Y and, therefore, if the 5 s genes were located here, one would expect to see a difference between wild-type males and females. (2) The right heterochromatic arm contains no known mutant genes, and at least part of it is dispensable (Cooper, 1959). Location in the other very small heterochromatic region is also rather improbable, but this possibility cannot be rigorously eliminated on the basis of the present experiments.

‡ The proportion of autosomal DNA is slightly different in mutant and wild-type flies, ranging from 77 to 82·5% of the total genome. In the present experiments this difference would be reflected by saturation plateaux differing by only 17 cts/min, and thus would not be readily detectable.

quantitative estimate of the maximum possible contribution of noise to the saturation value, we carried out the competition experiments detailed in Figure 6. It is seen that in the presence of as much as an 80-fold excess of competitor, there is an almost complete competition by the homologous 5 s species, but only a slight reduction by a heterologous species (4 s RNA). Thus the specificity is verified, and the amount of 5 s RNA binding which is not competed out by unlabeled 5 s RNA ($\sim 0.0012\%$) provides an upper limit for non-specific interactions. From these considerations it is concluded that about 0.0065 to 0.0077% of the DNA or 195 to 230 genes per haploid genome code for 5 s RNA (Table 2).

(e) *Number of genes for 4 S RNA*

Initial hybridization reactions with ^{3}H-labeled 4 s RNA also revealed the necessity of maintaining 100:1 ratio of unlabeled *Drosophila* 28+18 s RNA to ^{3}H-labeled 4 s RNA in order to eliminate a ^{3}H-labeled 28+18 s RNA contaminant. Under these conditions, 0.015% of the DNA hybridized to ^{3}H-labeled 4 s RNA as shown in Figure 7. The number of 4 s genes has been computed as indicated in Table 2.

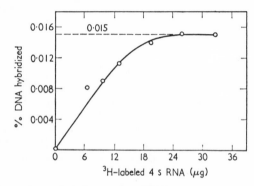

Fig. 7. Saturation of *Drosophila* DNA with ^{3}H-labeled 4 s RNA.
Filters containing 70 μg of DNA from adult Ore-R (♀ ♂) were incubated for 2 hr at 60°C in 3 ml. of 2 × SSC containing the indicated concentration ^{3}H-labeled 4 s RNA (77,500 cts/min/μg) and 100 times that amount of unlabeled 28 + 18 s RNA. Each point represents the mean of three independent determinations. At the saturation plateau the average filter measured approximately 950 cts/min with a background of 20 cts/min.

4. Discussion

Drosophila 5 s RNA is similar to the 5 s RNA's of many other species in having a $G+C:A + U$ ratio greater than one. This is particularly interesting in view of the distinctively low (G + C) composition of the 28 s and 18 s RNA's of *Drosophila*, and is consistent with the idea that throughout evolution the base composition of 5 s rRNA has diverged less than that of the 28 and 18 s rRNA's. Indeed, evidence has been accumulating which indicates that the primary structures of 5 s RNA from four mammalian species (human, mouse, rat and rabbit) are probably identical (Williamson & Brownlee, 1969) and the sequence similarities between *Escherichia coli* and human KB tumor cell 5 s RNA are quite striking (Madison, 1968; Forget & Weisman, 1969).

The DNA–RNA hybridization studies reported here demonstrated that in wild-type flies there are approximately 195 to 230 5 s RNA genes per haploid genome. Flies of the same stock also contain about 180 to 190 genes per haploid genome for both 28 and 18 s rRNA. Thus, in *Drosophila* the degree of redundancy for the various ribosomal RNA genes is roughly the same. The results of this investigation further indicate that few, if any, of the 5 s RNA genes are linked with the 28+18 s RNA genes on the sex chromosomes. Therefore, most, if not all, of the genes for 5 s RNA must reside on the autosomes. If the transcription of the 28 and 18 s genes is indeed co-ordinated with that of the 5 s genes, then this must occur by some mechanism which is not based on a linkage relationship.

Xenopus laevis is the only other eukaryote whose 5 s rRNA genes have been similarly analyzed (Brown & Weber, 1968). Although the question of linkage between 5 s and 28+18 s RNA genes was not completely resolved, the results indicated that in *Xenopus* the 5 s RNA genes are not intermingled with the 28 and 18 s RNA genes. From these same studies it was also concluded that whereas per haploid genome there are 450 copies of 28 s and 18 s RNA genes, there are at least 27,000 5 s RNA genes. This is in striking contrast to our finding in *Drosophila* and may represent an interesting example of biological diversity. However, it should be pointed out that there are substantial differences in methodology, as well as in the criteria used to evaluate the specificity of the DNA–5 s RNA hybrids in the *Xenopus* studies and in the experiments reported here, and that these could account for some of the discrepancy. When similar data are available for other eukaryotic species one may be able to clarify this point.

In *Bacillus subtilis* the 5 s RNA genes are tightly clustered with the 23 and 16 s RNA genes (Smith, Dubnau, Morell & Marmur, 1968). Kinetic evidence indicates that in this organism the precursor of 23 s RNA is converted to yield mature 23 s RNA and 5 s RNA (Hecht, Bleyman & Woese, 1968), and the fact that the synthesis of 23 and 5 s RNA is co-ordinately inhibited by actinomycin D is consistent with this hypothesis (Bleyman, Kondo, Hecht & Woese, 1969). Furthermore, in *E. coli*, levallorphan inhibits the production of all three rRNA's to the same extent (Roschenthaler, Devynck, Fromageot & Simon, 1969), and in *E. coli* (Rosset, Monier & Julien, 1964; Roschenthaler *et al.*, 1969), as well as in *B. subtilis* (Morell & Marmur, 1968), there appears to be no pool of free 5 s rRNA.

In eukaryotes the situation with respect to 5 s RNA is quite different. The present investigation has demonstrated the absence of clustering or even linkage between the 5 s and 28+18 s RNA genes in *Drosophila*. 20% of the cellular 5 s RNA exists as a pool in the nucleus in both HeLa (Knight & Darnell, 1967) and mouse L-cells (Perry & Kelley, 1968), and in L-cells the synthesis of 5 s RNA persists unabated when the transcription of 28 and 18 s RNA is inhibited by actinomycin D (Perry & Kelley, 1968). Data compatible with a nuclear pool of 5 s rRNA have also been reported for *Chironomous* salivary gland cells (Egyházi, Daneholt, Edström, Lambert & Ringborg, 1969; Edström & Daneholt, 1967).

These contrasting characteristics between prokaryotes and eukaryotes are consistent with the possibility that in prokaryotes the synthesis of all three ribosomal RNA components is co-ordinately regulated as a consequence of the clustering of their genes, but that with the evolution of nucleated cells the 5 s RNA genes and the 28 s + 18 s RNA genes were separated, and their interdependent control thereby lost.

The capable technical assistance of Joseph Gibbs, Dawn Kelley, Blanche Lewis, Roberta Ridley, Dana Tartof and Elizabeth Travaglini made this work a joy. Research was financially supported by U.S. Public Health Service postdoctoral fellowship grant to one of us

(K.D.T.) (CA-40014-01), grant GB-7051 from the National Science Foundation, grants CA-01613, CA-06927 and FR-05539 from the National Institutes of Health, and an appropriation from the Commonwealth of Pennsylvania.

REFERENCES

Aubert, M., Monier, R., Reynier, M. & Scott, J. F. (1967). *Proc. 4th Meeting FEBS*, p. 151. New York: Academic Press.

Bleyman, M., Kondo, M., Hecht, N. & Woese, C. (1969). *J. Bact.* **99**, 535.

Brown, D. D. & Weber, C. S. (1968). *J. Mol. Biol.* **34**, 661.

Cooper, K. W. (1959). *Chromosoma*, **10**, 535.

Edström, J. -E. & Daneholt, B. (1967). *J. Mol. Biol.* **28**, 331.

Egyházi, E., Daneholt, B., Edström, J. -E., Lambert, B. & Ringborg, U. (1969). *J. Mol. Biol.* **44**, 517.

Forget, B. G. & Weissman, S. M. (1969). *J. Biol. Chem.* **244**, 3148.

Greenberg, J. R. (1969). *J. Mol. Biol.* **46**, 85.

Hastings, J. R. B. & Kirby, K. S. (1966). *Biochem. J.* **100**, 532.

Hecht, N. B., Bleyman, M. & Woese, C. R. (1968). *Proc. Nat. Acad. Sci., Wash.* **59**, 1278.

Knight, E. & Darnell, J. E. (1967). *J. Mol. Biol.* **28**, 491.

Lindsley, D. L. & Grell, E. H. (1968). Carnegie Instn Publ. no. 627.

Loening, U. (1968). *J. Mol. Biol.* **38**, 355.

Madison, J. T. (1968). *Ann. Rev. Biochem.* **37**, 131.

Morell, P. & Marmur, J. (1968). *Biochemistry*, **7**, 1141.

Penman, S., Vesco, C. & Penman, M. (1968). *J. Mol. Biol.* **34**, 49.

Perry, R. P. & Kelley, D. E. (1968). *J. Cell Physiol.* **72**, 235.

Petrović, S. & Janković, V. (1962). *Bull. B. Kidrich Inst. Nucl. Sci.* **13**, 47.

Ritossa, F. M., Atwood, K. C., Lindsley, D. L. & Spiegelman, S. (1966). *Nat. Cancer Inst. Monog.* no. 23, p. 449.

Ritossa, F. M., Atwood, K. C. & Spiegelman, S. (1966a). *Genetics*, **54**, 663.

Ritossa, F. M., Atwood, K. C. & Spiegelman, S. (1966b). *Genetics*, **54**, 819.

Ritossa, F. M. & Spiegelman, S. (1965). *Proc. Nat. Acad. Sci., Wash.* **53**, 737.

Roschenthaler, R., Devynck, M. A., Fromageot, P. & Simon, E. J. (1969). *Biochim. biophys. Acta*, **182**, 481.

Rosset, R., Monier, R. & Julien, J. (1964). *Bull. Soc. Chim. Biol.* **46**, 87.

Smith, I., Dubnau, D., Morell, P. & Marmur, J. (1968). *J. Mol. Biol.* **33**, 123.

Weinberg, R. A., Loening, U., Willems, M. & Penman, S. (1967). *Proc. Nat. Acad. Sci., Wash.* **58**, 1088.

Williamson, R. & Brownlee, G. G. (1969). *FEBS Letters*, **3**, 306.

Processing of 45 s Nucleolar RNA

Robert A. Weinberg and Sheldon Penman

1. Introduction

Eucaryotic ribosomal RNA is synthesized and assembled into subunits in the nucleolus (Warner, 1966). In the mammalian nucleolus, the precursor to 18 s and 28 s rRNA is a 45 s molecule which is methylated and then cleaved to intermediate molecules which are eventually processed to become the 18 and 28 s RNA components of ribosomes (Perry, 1962; Scherrer & Darnell, 1962; Greenberg & Penman, 1966; Zimmerman & Holler, 1967). Previous work suggests that the 45 s is cleaved to 18 s rRNA plus a 32 s molecule, which is in turn converted to 28 s rRNA. The molecular weight of the 45 s is about twice the sum of the molecular weights of the 18 and 28 s rRNA. Other studies have shown that about half of the 45 s RNA is discarded in the course of forming the 18 and 28 s. This discarded portion is unmethylated (Weinberg, Loening, Willems & Penman, 1967) and has nucleotide sequences (Amaldi & Attardi, 1968; Jeanteur, Amaldi & Attardi, 1968; Jeanteur & Attardi, 1969) and base composition (Willems, Wagner, Laing & Penman, 1968) unlike the 18 and 28 s rRNA.

Besides the previously recognized 45, 32, 28 and 18 s species of nucleolar RNA, acrylamide gel electrophoresis of HeLa cell nucleolar RNA has revealed additional small peaks which correspond to RNA of approximately 41, 36, 24 and 20 s (Weinberg et al., 1967). The small amounts of these species normally present in isolated nucleoli precluded detailed analysis. However, this earlier work indicated that during infection, poliovirus interferes with normal nucleolar processing, and accumulation of the 41, 28, 20 and 18 s RNA's occurs. In the experiments discussed here, the poliovirus-induced accumulation of minor nucleolar RNA species is utilized to permit characterization of these minor species as intermediates in the conversion of 45 into 18 and 28 s rRNA.

95

The methylation of the 45 s RNA takes place entirely on the 45 s RNA molecule either during or immediately after synthesis (Greenberg, & Penman 1966). Virtually all the methyl groups found on the 45 s are found later on the 18 and 28 s molecules (Wagner, Penman & Ingram, 1967), and with one possible minor exception (Zimmerman, 1968), all methyl groups on the 18 and 28 s originate from methylation of the 45 s. The number of methyl groups differs in the 18 and 28 s rRNA molecules (Brown & Attardi, 1965; Iwanami & Brown, 1968). Thus, the number of methyl groups present on each nucleolar RNA molecule indicates unambiguously whether the molecule contains one or both regions destined to become ribosomal RNA. For example, the 32 s RNA will be shown to contain the same number of methyl groups as 28 s rRNA. This confirms that the 32 s is precursor to the 28 s alone, and that it cannot be a precursor to 18 s RNA. 45 s RNA will be shown to contain a number of methyl groups equal to the sum of methyl groups present on the 18 s and on the 28 s. This indicates that 45 s is precursor to both 18 and 28 s rRNA.

The molecular weight of the nucleolar RNA molecules must be known in order to determine the number of methyl groups per molecule. Molecular weights were determined in two different ways. One method depends upon the recently discovered semilogarithmic relationship between electrophoretic mobility and molecular weight on acrylamide gels (Loening, 1968; Bishop, Claybrook & Spiegelman, 1967; Peacock & Dingman, 1968). The other method depends upon the conservation of methyl groups throughout nucleolar RNA processing to calculate independently the molecular weight of nucleolar RNA species. These two methods of molecular weight measurement agree reasonably well and also agree with hydrodynamic measurements (McConkey & Hopkins, 1969).

Measurements of molecular weights and number of methyl groups are used to construct a processing scheme to explain the role of the newly discovered minor nucleolar species in the production of 18 and 28 s rRNA. This processing scheme indicates that two of the newly discovered minor RNA species are probably intermediates in the normal interconversion sequence of nucleolar RNA species while two others may represent aberrant processing, or artifactually produced RNA species.

2. Methods

The techniques used here have been described in detail in previously published reports. Nuclear isolation was performed as described by Penman (1966). Nucleolar isolation was described in a subsequent report (Penman, Vesco & Penman, 1968). The phenol extraction technique was first described by Scherrer & Darnell (1962), and was modified by the addition of chloroform (Penman, 1966) and the reduction of temperature from 65°C to 55°C (Wagner, Katz & Penman, 1967). Methionine labeling was performed in normal growth medium, supplemented with 2×10^{-5} M-adenosine and guanosine to reduce purine labeling (Weinberg et al., 1967). Polyacrylamide gel electrophoresis was performed as described by Loening (1967), as modified by Weinberg et al. (1967) and Weinberg & Penman (1968).

3. Results

(a) Methylation of nucleolar RNA species labeled during poliovirus infection

Poliovirus infection causes the accumulation of nucleolar RNA species normally present in small amounts in the nucleolus. The electropherogram of Figure 1(a) shows nucleolar RNA obtained from cells labeled during poliovirus infection. Cells were labeled with [methyl-³H]methionine, a methyl donor, and $Na_2H^{32}PO_4$ for three

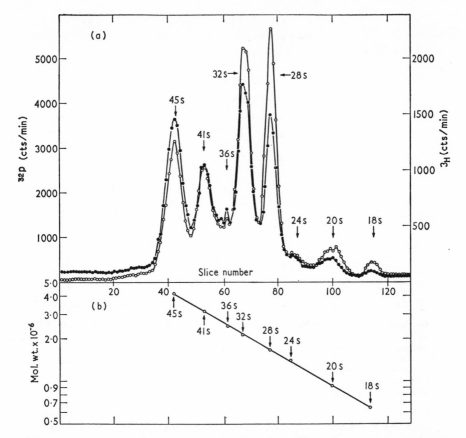

Fig. 1. Methionine and phosphate labeling of nucleolar RNA during poliovirus infection molecular weight calibration of nucleolar RNA's.

100 ml. of culture at 4×10^5 cells/ml. was centrifuged and the pellet resuspended in 50 ml. of fresh growth medium. The cells were labeled for 3 hr with 1 mc $^{32}PO_4$ and 1 mc of [*methyl*-^{3}H]methionine in the presence of 2×10^{-5} M-guanosine and 2×10^{-5} M-adenosine. The cells were then centrifuged and resuspended in 2 ml. of serum-free Eagle's medium containing 2 mM-guanidine and 4×10^9 plaque-forming units of poliovirus. Adsorption was allowed to proceed for 30 min at 7°C. The cells were centrifuged and then resuspended in the original labeling medium plus 2 mM-guanidine at 37°C for an additional 3 hr. —O—O—, ^{3}H, —●—●—, ^{32}P. (a) Gel electropherogram of nucleolar RNA run 9 hr on a 2·6% gel at 5 mA. (b) Logarithmic calibration of molecular weights of these RNA's *versus* electrophoretic mobility on this electropherogram. 18 and 28 s rRNA's were used as reference markers.

hours to achieve a relatively constant specific activity in acid-soluble precursor pools. The cells were then infected with poliovirus and harvested three hours after the beginning of poliovirus infection. Guanidine was present in the medium to prevent virulent growth of the virus.

The pattern shown in Figure 1(a) confirms that during poliovirus infection, aside from the 45 and 32 s RNA's normally present in large amounts in the nucleolar fraction, 41, 28, 20 and 18 s RNA also accumulate. The ratio of methyl groups (^{3}H: open circles) per nucleotide ($^{32}PO_4$: filled circles) increases greatly as the molecular weight of the nucleolar RNA intermediates decreases. Thus, the ^{3}H/^{32}P ratio of 18 s

RNA appears to be about 2·5 times higher than the $^3H/^{32}P$ ratio of 45 s RNA. The measured $^3H/^{32}P$ ratios of each of the nucleolar RNA species in this experiment are presented in Table 1.

TABLE 1

Methylation of nucleolar RNA species

s-value	Estimated mol. wt ($\times 10^{-6}$)	$\dfrac{[^3H]\text{Methyl}}{^{32}PO_4}$	Product of mol. wt $\times$ $^3H/^{32}P \times 10^{-4}$
45	4·1	0·363	149
41	3·1	0·448	140
32	2·1	0·485	102
28	1·65†	0·605	100
24	1·4	0·470	65
20	0·95	0·565	54
18	0·65‡	0·890	58
36	2·5		

These numbers refer to measurements of the molecular weights of the acidic forms of these RNA species.

† Petermann & Pavlovec, 1966.
 Hamilton, 1967.
 McConkey & Hopkins, 1969.
‡ Petermann & Pavlovec, 1966.
 McConkey & Hopkins, 1969.

Calculation of the number of methyl groups present on each of the nucleolar RNA species requires both these [^3H]methyl/$^{32}PO_4$ ratios, and a knowledge of the molecular weights of these RNA species. The logarithmic relationship between molecular weight and electrophoretic mobility has been established for a variety of RNA species (Loening, 1968; Bishop, Claybrook & Spiegelman, 1967; Peacock & Dingman, 1968). Using this relationship and the previously measured molecular weights of 18 and 28 s rRNA, the molecular weights of the nucleolar RNA species were estimated on the semi-logarithmic plot shown in Figure 1(b). The resultant molecular weights are shown in Table 1.

Multiplication of the $^3H/^{32}P$ (methyl groups/nucleotide) ratio of an RNA species by the estimated molecular weight of the RNA species gives a product which is proportional to the absolute number of methyl groups attached to the RNA species. These products are listed in the last column of Table 1. In each case, the product indicates the relative number of methyl groups present on each RNA species. Thus, Table 1 indicates that 28 s rRNA has about 100/58=1·7 times as many methyl groups as 18 s rRNA. In addition, the 18 s is 0·890/0·605=1·47 times as intensely methylated as the 28 s rRNA, in good agreement with the ratio of 1·4 of Iwanami & Brown (1968).

The RNA species listed in Table 1 are grouped into three classes, each class having a similar number of methyl groups. 45 and 41 s RNA contain a number of methyl groups about equal to the sum of methyl groups present on the 18 s plus those present on the 28 s. Since 41 s contains about the same number of methyl groups as 45 s, it too must be precursor to both 18 and 28 s. 32 s RNA, containing the same number of methyl groups as 28 s, is a precursor to 28 s alone. Both 24 and 20 s RNA contain about the same number of methyl groups as 18 s. Later discussion will indicate that the 20 s is probably a true precursor to 18 s rRNA, while the 24 s is an aberrant cleavage product, containing the 18 s region. The 36 s, whose $^3H/^{32}P$ ratio could not be estimated, may be a similarly aberrant cleavage product, discussed in detail below.

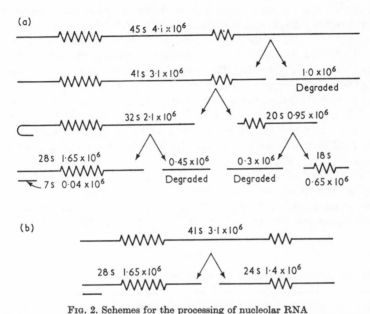

FIG. 2. Schemes for the processing of nucleolar RNA

(a) Proposed scheme for the normal cleavage sequence of 45 s into 18 and 28 s rRNA's. (b) Possible scheme accounting for the occasional appearance of 24 s RNA in the nucleolus. This scheme is supported by the methylation data of Table 1.

Figure 2(a) presents a scheme for the cleavage of 45 s ribosomal RNA precursor, as deduced from the data of Table 1. The molecular weights in the Figure are from Table 1. The zig-zag portions indicate methylated regions of the RNA molecules. We have no knowledge of the topological order of the different regions of the 45 s molecule and the order indicated in this Figure is arbitrary. In addition to the RNA species discussed here, the 7 s ribosomal RNA is included in this Figure. This RNA, discovered and characterized by Pène, Knight & Darnell (1968), is about 150 nucleotides long, is hydrogen-bonded to the 28 s rRNA in the ribosome, and is produced simultaneously with the cleavage of 32 s into 28 s rRNA. Many experiments, not discussed here, indicate that this 7 s is also derived from the 32 s (Pène *et al.*, 1968; Weinberg, unpublished observations). The 5 s rRNA is not derived from any of the RNA species shown in this Figure (Brown, 1968; Perry & Kelley, 1970).

(b) *Independent measurement of the molecular weight of 32 and 45 s*

The methylation and molecular weights shown in Figure 2(a) are consistent with the data presented in Table 1. Neither the 24 nor 36 s RNA seen in Figure 1(a) is included in this scheme because their molecular weights are incompatible with their integration into this or similar processing schemes containing the 41, 32 and 20 s. In addition, the 24 and 36 s are present in small or non-existent amounts in normal preparations and their amounts are not enhanced during poliovirus infection. The 24 and 36 s RNA's appear consistent only with a processing scheme in which the normal sequence of cleavages is altered (Fig. 2(b)).

The exclusion of 36 and 24 s RNA from the proposed processing scheme depends critically upon the accuracy of the molecular weight determinations shown in Table 1. An experiment was therefore performed to obtain an independent measure of the molecular weights of 28, 32 and 45 s relative to 18 s RNA. The first part of this experiment provides an independent measurement of the ratio of molecular weights of 28 to 18 s rRNA. This ratio provides an independent determination of the slope of the calibration curve of Figure 1(b). The second part of the experiment assumes that methyl groups are conserved during processing of 45 to 18 s and 28 s rRNA's. By measuring the $[^3H]$methyl/$^{32}PO_4$ ratio of 45, 32, 28 and 18 s RNA's from long-labeled cells, one can calculate values for the molecular weights of 45 and 32 s rRNA relative to the molecular weights of 18 and 28 s rRNA's. These experiments serve to establish that no conformation change significantly affects the molecular weight determination by gel electrophoresis.

Cells were labeled for 48 hours in medium containing both $^{32}PO_4$ and $[methyl$-$^3H]$-methionine. The RNA from a sample of the cytoplasmic extract was deproteinated by adding sodium dodecyl sulfate and EDTA, and a small portion examined by gel electrophoresis. The result is shown in Figure 3(a). The ratio of radioactivity

TABLE 2

Molecular weight determination of nucleolar RNA from $[^3H]$methyl/$^{32}\dot{P}O_4$ ratios

Molecular weight estimate of 32 and 45 s from 48 hr continuous $^{32}PO_4$ and $[methyl$-$^3H]$methionine label.

$$\frac{^{32}PO_4 \text{ cts/min of cytoplasmic 28 s rRNA}}{^{32}PO_4 \text{ cts/min of cytoplasmic 18 s rRNA}} = 2\cdot49$$

Assumed molecular weight of 18 s $= 0\cdot65 \times 10^6$ daltons (Petermann & Pavlovec, 1966; Hamilton, 1967). Calculated molecular weight of 28 s $= 0\cdot65 \times 10^6 \times 2\cdot49 = 1\cdot62 \times 10^6$ daltons.

RNA species	$\dfrac{[^3H]\text{methyl cts/min}}{^{32}PO_4 \text{ cts/min}}$	Calculated mol. wt (daltons $\times 10^{-6}$)	Product of mol. wt $\times$ $^3H/^{32}P$ ratio
18 (cytoplasmic)	0·85	0·65	0·55
28 (cytoplasmic)	0·54	1·62	0·87

Calculated molecular weight of 32 s (nucleolar):
$^3H/^{32}PO_4$ of 32 s $= 0\cdot418$

Assuming that 32 s contains the same number of methyl groups as 28 s, molecular weight of 32 s $= 0\cdot87/0\cdot418 = 2\cdot1 \times 10^6$ daltons.

Calculated molecular weight of 45 s (nucleolar):
$^3H/^{32}PO_4$ of 45 s $= 0\cdot321$

Assuming that 45 s contains the sum of methyl groups of 18 and 28 s, the molecular weight of
$$45 \text{ s} = \frac{(0\cdot55 + 0\cdot87)}{0\cdot321} = 4\cdot4 \text{ million daltons}$$

100

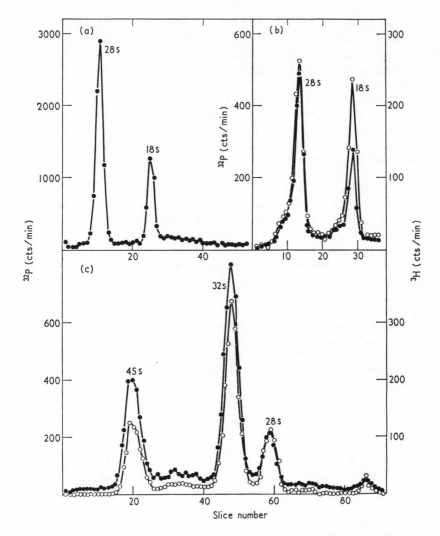

Fig. 3. RNA extracted from cells labeled 48 hr with ³²PO₄ and [*methyl*-³H]methionine.

100 ml. of culture containing 1×10^5 cells/ml. was grown 48 hr in the presence of 0·5 mc of ³²PO₄, 0·5 mc of [*methyl*-³H]methionine, 5×10^{-5} M-adenosine, and 5×10^{-5} M-guanosine. The cells were fractionated into cytoplasmic, nucleoplasmic, and nucleolar fractions.

(a) Cytoplasmic RNA. 20 μl. (1%) of the first cytoplasmic supernatant was applied directly onto a 2·6% acrylamide gel after addition of sodium dodecyl sulfate to 0·5% and EDTA to 0·01 M. There was no prior phenol extraction. Only the ³²P profile is shown. The gel was run for 3 hr.

(b) Cytoplasmic RNA—phenol extracted. 10% of the cytoplasmic RNA was phenol-extracted at 25°C. Electrophoresis was as above.

(c) Nucleolar RNA—phenol extracted. The nucleolar RNA was phenol-extracted at 55°C. One-half of the nucleolar RNA was applied to the gel shown here. Since electropherograms of samples of the nucleolar RNA showed a high ³²P/³H ratio, the RNA shown in (b) and (c) was stored in 70% ethanol at −20°C for 3 weeks before electrophoresis to allow decay of the ³²PO₄, minimizing the spillover. The 2·6% gel was run 9 hr at 5 mA. —●—●—, ³²P; —○—○—, ³H.

(^{32}P) in 28 s to that in 18 s RNA should be proportional to the ratio of their molecular weights. The length of the labeling period was sufficient to minimize errors arising from the lag in the time of entry into the cytoplasm of 28 s RNA compared to 18 s RNA, and from transient differences in the specific activities of the nucleotide pools. Also the technique of gel electrophoresis does not suffer from the geometrical limitations of sucrose gradients which result in a recovery dependent on the distance a molecule has sedimented. Since deproteinization with phenol is omitted, the possibility of selective recovery of either 18 or 28 s RNA is avoided. No radioactive material was trapped at the top of the gel. The radioactivity measured is presented in Table 2. The measured ratio of radioactivity is 2·49 which is in excellent agreement with the published value of 2·54 (Petermann & Pavlovec, 1966).

The molecular weight of 45 and 32 s RNA was estimated by comparing the [^{3}H]methyl/^{32}PO$_4$ present in the precursor species with that present in the mature cytoplasmic species, assuming an absolute conservation of methyl groups. In principle the data from Figure 1 could be used in this calculation. However, the accuracy of the [^{3}H]methyl/^{32}PO$_4$ ratio of 18 s RNA was limited by the small amount of 18 s RNA present in the nucleolus. In addition, although methylation of RNA during virus infection is not grossly aberrant, small changes could conceivably affect a molecular weight determination.

Cytoplasmic and nucleolar RNA of the 48-hour labeled cells were extracted with phenol and analyzed by gel electrophoresis. The electropherograms are shown in Figure 3(b) and (c) and the results summarized in Table 2.

The molecular weight estimated for 32 s is $2 \cdot 1 \times 10^6$ and for 45 s, $4 \cdot 4 \times 10^6$ daltons. Considering the limitations of the method, this result is in reasonable agreement with the previous estimate from electrophoretic mobility.

4. Discussion

Previous experiments have established the presence of 45 s rRNA precursor, 32 s RNA intermediate, and 28 and 18 s rRNA's in the nucleolus. Analysis of gel electropherograms showed also the presence of 41, 36, 24 and 20 s RNA's (Weinberg et al., 1967). The present experiments provide evidence for the processing scheme shown in Figure 2(a).

The conclusions illustrated in Figure 2(a) depend upon the accuracy of the molecular weight determination. There are two possible sources of error in the molecular weight determination by acrylamide gel. A significant change in conformation between 45 and 18 and 28 s molecules could lead to deviation from the assumed logarithmic relationship between molecular weight and electrophoretic mobility. Another source of error might arise from inaccuracy in the previously determined molecular weights of 18 and 28 s RNA. The molecular weight scale shown in Figure 1 is constructed by extrapolation from the cited published values for the molecular weights of 18 and 28 s RNA. The accuracy of the scale is particularly sensitive to an error in the ratio of molecular weights of 28 s compared to 18 s, since the slope of the calibration is determined by this ratio. The 48-hour labeling experiment provides independent confirmation of the ratio of molecular weights of 28 to 18 s, and of the molecular weights of 32 and 45 s.

Figure 2(a) illustrates what are probably the usual steps in the processing of 45 s RNA. The molecular weight determinations reported here strongly suggest that 36

102

and 24 s are not normal intermediates in this scheme. If 36 s were placed as intermediate between 41 and 32 s, then the cleavage of 41 s would result in a 36 and a 20 s molecule. The molecular weight of 41 s would then have to be at least 3.5×10^6 daltons and the molecular wieght of the 45 s, by extrapolation, would be raised to about 4.7×10^6 daltons. This high molecular weight probably exceeds the uncertainties of the molecular weight determinations performed here. However, the accuracy of present methods cannot absolutely preclude 36 s as a normal intermediate.

It should be noted that it is possible to generate 36 and 24 s RNA without violating the scheme of Figure 2(a) by changing the order of the cleavage sequence. For example, 41 s RNA might be cleaved to produce a normal 28 and a 24 s molecule as shown in Figure 2(b). Similarly, 41 s might be cleaved to produce a normal 18 and 36 s molecule. These altered sequences might arise from two sources: *in vivo* the sequence of the cleavages may occasionally vary. Alternatively, during isolation of the nucleoli, certain cleavages may take place *in vitro* out of the usual order (Vesco & Penman, 1968; Liau, Craig & Perry, 1968).

Almost one-half of the original 45 s molecule is not utilized to make 18 and 28 s rRNA's. The discarded fragments are extremely high in their G + C content (Willems *et al.*, 1968) and are probably completely unmethylated. We have been unable to detect any trace of the discarded fragments *in vivo*. They are metabolically highly unstable. A lifetime of as short as 15 minutes would have permitted their detection. They are probably rapidly degraded after their cleavage from the methylated portions of the ribosomal precursor.

Some of the minor nucleolar intermediates discussed here have recently been reported in other tissues. Markov & Emanuiloff (1968) have found RNA species corresponding to 41, 24 and 20 s in ascites tumor cells. Edstrom & Daneholt (1967) have found a 20 s precursor to 18 s rRNA in insect nucleoli. Vesco, in this laboratory, has demonstrated a pronounced 20 s RNA intermediate in cultured *Xenopus* cells (personal communication).

This work was supported by awards CA08416-03 from the National Institutes of Health and GB5809 from the National Science Foundation. One of us (S.P.) is a Career Development Awardee of the U.S. Public Health Service, GM16127-03. The other (R.A.W.) was a pre-doctoral fellow of the National Institutes of Health, grant F1-GM-23, 898-03. The capable assistance of Deana Fowler and Elizabeth Loutrel is gratefully acknowledged.

REFERENCES

Amaldi, F. & Attardi, G. (1968). *J. Mol. Biol.* **33**, 737.
Bishop, D. H. L., Claybrook, J. R. & Spiegelman, S. (1967). *J. Mol. Biol.* **26**, 373.
Brown, D. D. (1968). *Current Topics in Developmental Biol.* **2**, 47.
Brown, G. M. & Attardi, G. (1965). *Biochem. Biophys. Res. Comm.* **20**, 298.
Edstrom, J. E. & Daneholt, B. (1967). *J. Mol. Biol.* **28**, 331.
Greenberg, H & Penman, S. (1966). *J. Mol. Biol.* **21**, 527.
Hamilton, M. G. (1967). *Biochim. biophys. Acta*, **134**, 473.
Iwanami, Y. & Brown, G. M. (1968). *Arch. Biochem. Biophys.* **126**, 8.
Jeanteur, P., Amaldi, F. & Attardi, G. (1968). *J. Mol. Biol.* **33**, 747.
Jeanteur, Ph. & Attardi, G. (1969). *J. Mol. Biol.* **45**, 305.
Liau, M. C., Craig, N. C. & Perry, R. P. (1968). *Biochim. biophys. Acta*, **169**, 196.
Loening, U. (1967). *Biochem. J.* **102**, 251.
Loening, U. (1968). *Biochem. J.* **113**, 131.

Markov, G. G. & Emanuiloff, E. A. (1968). *Abhandlung der Deutschen Akademie der Wissenschaften zu Berlin*, International Symposium at Castle Reinhardsbrunn, p. 323.

McConkey, E. H. & Hopkins, J. W. (1969). *J. Mol. Biol.* **39**, 545.

Peacock, A. C. & Dingman, C. W. (1968). *Biochemistry*, **7**, 668.

Pène, J. J., Knight, E., Jr. & Darnell, J. E., Jr. (1968). *J. Mol. Biol.* **33**, 609.

Penman, S. (1966). *J. Mol. Biol.* **17**, 117.

Penman, S., Vesco, C. & Penman, M. (1968). *J. Mol. Biol.* **34**, 49.

Perry, R. P. (1962). *Proc. Nat. Acad. Sci., Wash.* **48**, 2179.

Perry, R. P. & Kelley, D. E. (1970). *J. Cell Phys.* in the press.

Petermann, M. L. & Pavlovec, A. (1966). *Biochim. biophys. Acta*, **114**, 264.

Scherrer, K. & Darnell, J. (1962). *Biochem Biophys. Res. Comm.* **7**, 486.

Vesco, C. & Penman, S. (1968). *Biochim. biophys. Acta*, **169**, 188.

Wagner, E. K., Katz, L. & Penman, S. (1967). *Biochem. Biophys. Res. Comm.* **28**, 152.

Wagner, E., Penman, S. & Ingram, V. (1967). *J. Mol. Biol.* **29**, 371.

Warner, J. R. (1966). *J. Mol. Biol.* **19**, 383.

Weinberg, R. A., Loening, U., Willems, M. & Penman, S. (1967). *Proc. Nat. Acad. Sci., Wash.* **58**, 1088.

Weinberg, R. A. & Penman, S. (1968). *J. Mol. Biol.* **38**, 289.

Willems, M., Wagner, E., Laing, R. & Penman, S. (1968). *J. Mol. Biol.* **32**, 211.

Zimmerman, E. & Holler, B. (1967). *J. Mol. Biol.* **23**, 149.

Zimmerman, E. F. (1968). *Biochemistry*, **7**, 3156.

Membrane-bound Ribosomes in HeLa Cells

I. Their Proportion to Total Cell Ribosomes and their Association with Messenger RNA

BARBARA ATTARDI, BARBARA CRAVIOTO AND GIUSEPPE ATTARDI

1. Introduction

In all animal cells ribosomes occur in two different topographical situations, namely, either attached to membranes of the endoplasmic reticulum or free in the cytoplasmic matrix. The proportion of bound and free ribosomes varies in different types of cells: in cells which are specialized for the synthesis of protein destined to be exported, like liver or pancreas, the major part of ribosomes are associated with the endoplasmic reticulum (Palade, 1956,1958); on the contrary, in rapidly multiplying cells, such as those in embryonic tissues or those growing *in vitro*, the great majority of ribosomes are free (see review by Porter, 1961). In addition to these two groups, the existence of a minor group of ribosomes in mitochondria has been postulated in animal, as in other eukaryotic cells, on the basis of direct electron microscopic and biochemical observations (Rendi & Warner, 1960; Truman, 1963; Watson & Aldridge, 1964; Swift, 1965; André & Marinozzi, 1965; Elaev, 1966; O'Brien & Kalf, 1967a,b; Dubin & Brown, 1967) and of indirect evidence bearing on the protein synthesizing capacity of these organelles (see, among others, Roodyn, Reis & Work, 1961; Roodyn, Suttie & Work, 1962; Truman & Korner, 1962; Kroon, 1963a,b,c; Wheeldon & Lehninger, 1966); the occurrence of intramitochondrial ribosomes with distinctive rRNA components has been reported in *Neurospora* (Küntzel & Noll, 1967; Rifkin, Wood &

Luck, 1967; Dure, Epler & Barnett, 1967) and yeast (Rogers, Preston, Titchener & Linnane, 1967; Wintersberger, 1967).

The ribosomes bound to endoplasmic reticulum in secretory cells have been the object of numerous investigations concerning their involvement in the synthesis and transport of protein (Siekevitz & Palade, 1960; Redman, Siekevitz & Palade, 1966; Henshaw, Bojarski & Hiatt, 1963; Howell, Loeb & Tomkins, 1964), their mode of attachment to the membranes (Sabatini, Tashiro & Palade, 1966; Blobel & Potter, 1967b), and their relationship with free ribosomes (Moulé, Rouiller & Chauveau, 1960; Webb, Blobel & Potter, 1964; Loeb, Howell & Tomkins, 1965,1967; Cammarano, Giudice & Lukes, 1965; Manganiello & Phillips, 1965). The membrane-bound ribosomes of animal cells growing *in vitro*, though recognized by electron microscopists (Epstein, 1961; Journey & Goldstein, 1961; Fuse, Price & Carpenter, 1963), have, on the contrary, been disregarded in most biochemical investigations. The occurrence of bound ribosomes in these cells, which lack, in general, an obvious secretory activity, suggests that the attachment of ribosomes to membranes is not exclusively related to the synthesis of proteins to be exported. The possibility that membrane-bound polysomes may be involved in the synthesis of membrane proteins is suggested by observations made in differentiating hepatocytes of newborn rats (Dallner, Siekevitz & Palade, 1966). Animal cells growing *in vitro* are favorable material for the study of the functional role not immediately related to secretion of the attachment of ribosomes to membranes. The association with membranes of an appreciable fraction of ribosomes and mRNA in these rapidly multiplying cells has also relevance for the study of the mRNA metabolism and of the dynamics of polysome assembly and function. Evidence suggesting that polysomes of the rough endoplasmic reticulum in HeLa cells contain mRNA of mitochondrial origin has been recently reported (Attardi & Attardi, 1968). As a preliminary to the study of the physiological significance of membrane-bound ribosomes in HeLa cells, in particular, of their possible involvement in membrane protein synthesis, an electron microscopic and biochemical investigation has been carried out on these ribosomes, with special regard to their proportion to total cell ribosomes, their attachment to the membranes and their association with mRNA. It has been found that, in these cells, from 10 to 15% (and possibly as many as 20%) of the ribosomes are attached to elements of the endoplasmic reticulum. The majority of these membrane-bound ribosomes (65 to 70%) are recovered as polysomes after sodium deoxycholate treatment.

2. Materials and Methods

(a) Cells and method of growth

The method of growth of HeLa cells has been previously described (Amaldi & Attardi, 1968). The cultures used here were free of any detectable contamination by pleuropneumonia-like organisms (Mycoplasma).

(b) Buffers

The buffer designations are: (1) T: 0·01 M-Tris buffer (pH 7·1); (2) TM: 0·01 M-Tris buffer (pH 7·1), 0·00015 M-MgCl$_2$; (3) TKM: 0·01 M-Tris buffer (pH 7·1), 0·01 M-KCl, 0·00015 M-MgCl$_2$; (4) SMET (Parsons, Williams & Chance, 1966): 0·07 M-sucrose, 0·21 M-D-mannitol, 0·001 M-Tris buffer (pH 7·1), 0·0001 M-EDTA; (5) TKV: 0·05 M-Tris buffer (pH 7·1), 0·025 M-KCl, 0·001 M-EDTA; (6) low ionic strength TKV: 0·01 M-Tris buffer (pH 7·1), 0·01 M-KCl, 0·001 M-EDTA; (7) acetate–NaCl buffer: 0·01 M-acetate buffer (pH 5·0), 0·1 M-NaCl; (8) sodium dodecyl sulfate buffer (Gilbert, 1963): 0·01 M-Tris buffer (pH 7·0), 0·1 M-NaCl, 0·001 M-EDTA, 0·5% sodium dodecyl sulfate.

(c) Labeling conditions

Pulse labeling of RNA was carried out by exposing exponentially growing HeLa cells (2 to 3 × 10⁵ cells/ml.) for various times to [5-³H]uridine (17·3 to 28·8 mc/μmole, 0·3 to 10·0 μc/ml.) or [2-¹⁴C]uridine (30 to 52 μc/μmole, 0·025 to 0·07 μc/ml.) in modified Eagle's medium with 5% dialyzed calf serum. Long-term labeling of RNA was carried out by growing cells for 24 to 26 hr in the presence of [5-³H]uridine (0·3 to 1·25 μc/ml.), unless otherwise specified. DNA was labeled by growing cells for 24 hr in the presence of [³H-*methyl*]thymidine (22·6 mc/μmole, 0·25 μc/ml.).

(d) Preparation and analysis of subcellular fractions

All operations described below were carried out at 2 to 3°C. The labeled cells were washed three times with 0·13 M-NaCl, 0·005 M-KCl, 0·001 M-MgCl₂ and then resuspended in 6 vol. TKM. After 2 min, the suspension was homogenized with an A. H. Thomas homogenizer (motor-driven pestle, ∼1600 rev./min, 8 to 10 strokes). Under these conditions of homogenization, at most 60 to 70% of the cells were broken: these relatively mild conditions of cell breakage were chosen so as to minimize the rupture of nuclei and the subsequent release of labeled nuclear RNA components. (Less than 1% of the total cell [³H]thymidine-labeled DNA was found in the cytoplasmic fraction under these conditions.) After addition of sucrose to 0·25 M, the homogenate was centrifuged at 1160 g_{av} for 3 min to sediment nuclei, unbroken cells and large cytoplasmic debris. The supernatant (*total cytoplasmic fraction*) was spun at 8100 g_{av} for 10 min; the pellet thus obtained (including any loose fluffy layer) was resuspended in 0·25 M-sucrose in TM (one-half of the volume of the homogenate) and, after a spin at 1100 g_{av} for 2 min to sediment any residual nuclei, recentrifuged at 8100 g_{av} for 10 min. The pellet (and any fluffy layer) was resuspended in 0·25 M-sucrose in T buffer (1·0 or 2·0 ml.), for material deriving from 1·0 to 2·5 × 10⁸ cells) (Mg ions were omitted at this stage and in the following steps aimed at fractionation of the intact membrane components, in order to reduce the possibility of aggregation): this represented the 8100 g *membrane fraction*, which contained the bulk of mitochondria and of elements of rough endoplasmic reticulum, in addition to smooth membrane components. The first 8100 g supernatant was centrifuged at 15,800 g_{av} for 20 min to separate a small amount of slower sedimenting mitochondria and other membrane elements from the supernatant fraction containing the great majority of *free polysomes* and all free monomers and "native" ribosomal subunits.

Buoyant density fractionation of the 8100 g membrane components was carried out by centrifugation through a 30 to 48% (w/w) sucrose gradient in T buffer in the Spinco SW 25·1 rotor for 18 to 20 hr at 25,000 rev./min. For the analysis of the membrane-associated polysomes, the 8100 g membrane fraction was treated with 1% NaDOC† and centrifuged through a 15 to 30% (w/w) sucrose gradient in TKM (25 ml., prepared above 3 ml. of 64% (w/w) sucrose in TKM) in the SW25·1 rotor for 90 to 110 min at 24,000 rev./min; in some experiments the NaDOC-lysed membrane fraction was treated with EDTA (10⁻³ to 10⁻² M), and centrifuged through a 15 to 30% sucrose gradient (over 3 ml. of 64% sucrose) in low ionic strength TKV at the speed and for the time indicated above. For the study of the effect of EDTA on the untreated 8100 g membrane fraction, a suspension of this in 0·25 M-sucrose in T buffer (1·0 to 2·5 ml. for material deriving from about 1·3 × 10⁸ cells) was brought to 3 × 10⁻² M-EDTA, kept in the cold for 10 min, and then centrifuged at 11,000 g_{av} for 10 min. The supernatant fraction was carefully sucked up; the pellet was rinsed with 0·5 ml. of 0·25 M-sucrose in T buffer containing 1·5 × 10⁻² M-EDTA (which was pooled with the supernatant fraction) and resuspended in 4·0 ml. of the same medium: the suspension was immediately recentrifuged at 11,000 g_{av} for 10 min. The final pellet was resuspended in 1·0 or 2·0 ml. of 0·25 M-sucrose in T buffer, and either run on a 30 to 48% sucrose gradient in the same buffer (see above), or treated with 1% NaDOC and run on a 15 to 30% sucrose gradient (over 3 ml. of 64% sucrose) in TKV for 8 hr at 25,000 rev./min. Isolation of the ribosomal subunits released from the membrane fraction by EDTA treatment was carried out by centrifuging the EDTA supernatant fraction through a 15 to 30% sucrose gradient in TKV in the SW25·2 rotor for 17 hr at 25,000 rev./min.

† Abbreviations used: NaDOC, sodium deoxycholate; rRNA, ribosomal RNA; mRNA, messenger RNA; tRNA, transfer RNA.

Separation of free polysomes from monomers and native ribosomal subunits and from soluble components was carried out by centrifuging 3 to 5 ml. of the 15,800 g supernatant of the total cytoplasmic fraction through a sucrose gradient consisting, from the meniscus to the bottom, of 6 ml. 23% (w/w) sucrose, 12 ml. 23 to 55% sucrose gradient, and 6 ml. 55% sucrose, all in TKM (SW25·1 rotor, 25,000 rev./min, 5 hr). For better resolution of free monomers and native ribosomal subunits, the 15,800 g supernatant fraction was centrifuged through a 15 to 30% sucrose gradient in TKM for 8 hr at 25,000 rev./min. "Derived" ribosomal subunits were obtained from the free polysome–monomer fraction (pelleted by centrifuging the 15,800 g supernatant fraction at 105,000 g_{av} for 90 min) by treatment for 10 min with 10^{-2} M-EDTA, and separated on a 15 to 30% sucrose gradient in TKV as described above for the subunits from the 8100 g membrane fraction.

(e) Extraction and analysis of RNA

RNA was generally released from the membrane components and their NaDOC lysis products by treatment with 1% sodium dodecyl sulfate and from free polysomes with 0·5% sodium dodecyl sulfate, precipitated with 2 vol. of ethanol in the presence of 0·1 M-NaCl, dissolved in sodium dodecyl sulfate buffer, and run through a 15 to 30% (w/w) sucrose gradient (over 3 ml. of 64% sucrose) in sodium dodecyl sulfate buffer in the SW 25·1 rotor for 14 hr at 20,000 rev./min, 20°C.

For the analysis of radioactivity, a portion of each fraction was precipitated in the cold with 15% trichloroacetic acid by using 200 μg bovine serum albumin as a carrier, and the precipitate collected on a Millipore membrane. The isotope-counting procedures have been described elsewhere (Attardi, Parnas, Hwang & Attardi, 1966).

(f) Cytochrome oxidase assay

Cytochrome oxidase assay was carried out by a modification of the procedure of Smith (1954). 0·1-ml. portions of the sucrose gradient fractions were mixed in a cuvette with 2·9 ml. 18 μM-solution of reduced cytochrome c in 0·04 M-PO$_4$ buffer, pH 6·2. The decrease of O.D. at 550 mμ at 25°C was measured at 10-sec intervals for 4 to 6 min.

(g) Electron microscopy

HeLa cells were fixed, either in suspension or as a pellet, in sodium acetate–barbital buffer (pH 7·4–1% OsO$_4$ (Palade, 1952) for 1·5 hr at 2 to 3°C. For the preparation and fractionation of membrane components to be utilized for electron microscopy, the procedure described in Materials and Methods (d) was used, with the difference that the buoyant-density centrifugation was carried out in 30 to 48% sucrose gradient in 0·01 M-phosphate buffer, pH 7·0, to avoid interference in fixation by Tris buffer (Parsons et al., 1966): the components corresponding to different portions of the buoyant-density pattern were diluted 4 times with 0·01 M-phosphate buffer, pH 7·0, and pelleted by centrifugation in the Spinco 40 rotor for 60 min at 20,000 rev./min; the pellets were then fixed with sodium acetate–barbital buffer–1% OsO$_4$ for 1 hr at 2 to 3°C. In all cases, after dehydration with a graded series of ethyl alcohols, the embedding was made in Araldite 502 (CIBA Company, Inc., Kimberton, Pa.). Sections 600 to 900 Å thick were cut with glass knives on an LKB ultrotome and stained with uranyl acetate (64% in methyl alcohol, 20 min) and lead citrate (0·4% adjusted to pH 12, 2 to 4 min) (modified from Reynolds, 1963). The specimens were examined in a Philips electron microscope.

3. Results

(a) Electron microscopy of sections of HeLa cells

An electron microscopic examination of thin sections of the HeLa cells used in the present study has shown the presence of numerous elements of both rough and smooth endoplasmic reticulum scattered throughout the cytoplasm (Plate I(a)). The rough elements appear as sections, at various angles, of tubules and vesicles of different size and shape and of short cisternae. Where the rough elements are cut tangentially,

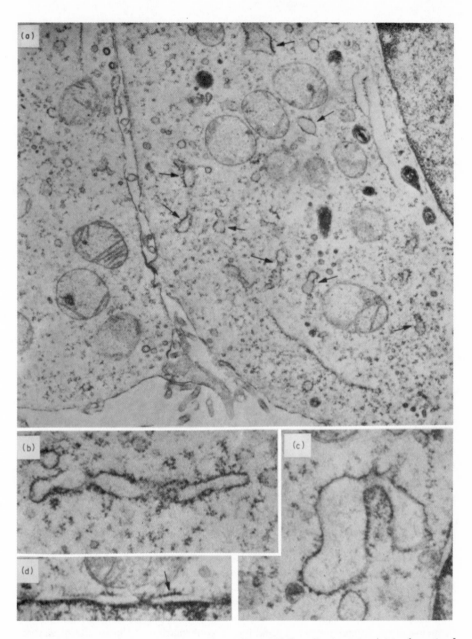

PLATE I. (a) Portion of the cytoplasm of two adjacent HeLa cells. Arrows point to elements of rough endoplasmic reticulum. Note the abundance of free polysomes in the cytoplasmic matrix. × 16,250.

(b) Cisterna of rough endoplasmic reticulum. Note rows of ribosomes along the edges and a rosette in the right half, at a point where the limiting membrane has been cut tangentially. × 35,000.

(c) Branched cisterna of rough endoplasmic reticulum. × 35,000.

(d) A row of ribosomes attached to the outer nuclear membrane. × 35,000.

one can see ribosomes arranged in rows, spirals or rosettes (Plate I(b)). The outer nuclear membrane sometimes shows attached ribosomes (Plate I(d)). Most of the rough elements appear to be isolated; sometimes, however, they are grouped and may also be in communication (Plate I(c)).

Most ribosomes in HeLa cells, as in other rapidly multiplying cells, are free in the cytoplasm, mainly in the form of aggregates (polysomes) of various size (Plate I(a)).

(b) *Fractionation of the 8100* g *membrane components*

The 8100 **g** membrane fraction contains, as was mentioned earlier, the bulk of mitochondria and elements of rough endoplasmic reticulum, in addition to smooth membrane components. The latter could be separated from rough endoplasmic reticulum and mitochondria on the basis of buoyant-density centrifugation in sucrose gradient.

A large number of experiments were performed to try to obtain a satisfactory resolution of rough endoplasmic reticulum from mitochondria by differential centrifugation or by sedimentation velocity or buoyant-density fractionation in a sucrose gradient. For these experiments, a variety of media was utilized both for homogenization (TKM, low ionic strength TKV, SMET) and sucrose gradients (T buffer, with or without addition of EDTA or CsCl (according to an adaptation of the procedure of Dallner *et al.* (1966)). These experiments gave disappointing results. At best, a partial separation of the two types of organelles was obtained: this was due to the extensive overlapping in their sedimentation properties (Attardi & Attardi, 1968) and density (Fig. 1). Figure 1 shows the results of a typical buoyant-density centrifugation of the 8100 **g** membrane fraction in a 30 to 48% sucrose gradient in T buffer. The $O.D._{260}$ analysis reveals a broad band occupying the region of the gradient corresponding to ρ values from 1·16 to 1·20 g/ml. and a smaller band centered around $\rho \sim 1·14$ g/ml. (It should be pointed out that the $O.D._{260}$ of membrane components only in part represents true absorption, in part being caused by light scattering.) The main band contains mainly mitochondria and elements of rough endoplasmic reticulum, as revealed by the cytochrome oxidase assay and by electron microscopic examination of sections of the pelleted components; in the light band, on the other hand, the electron microscopic analysis reveals smooth membrane structures. Both after short and long exposure of the cells to labeled RNA precursors, the structures containing the newly synthesized RNA are found in the region of the main $O.D._{260}$ band; very little labeled RNA appears in correspondence with the band of smooth membrane components. It was previously shown (Attardi & Attardi, 1968) that after a very short pulse (3 min) with [^{3}H]uridine, the majority of the newly synthesized RNA in the membrane fraction, if not all, is intramitochondrial; with increasing pulse length or after a pulse–chase, an increasing proportion of labeled RNA is found to be associated with extramitochondrial structures, presumably elements of rough endoplasmic reticulum. In confirmation of the results of electron microscopic examination, the buoyant-density distribution in a sucrose gradient of the structures containing three-minute [^{3}H]uridine-labeled RNA overlaps extensively that of the structures containing RNA labeled during a two-hour pulse with [^{14}C]-uridine (Fig. 1), or during a 24- to 48-hour exposure to the precursor, which are in their great majority elements of rough endoplasmic reticulum (see below). Since the aim of the present work was to obtain the over-all picture of the properties of the endoplasmic reticulum-bound ribosomes in HeLa cells and to try to estimate their

110

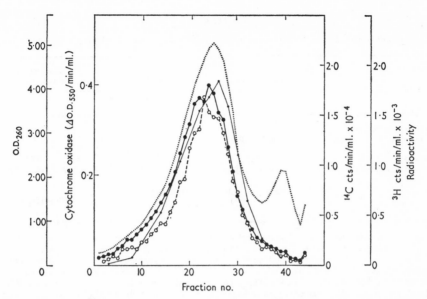

Fig. 1. Buoyant density distribution in sucrose gradient of components of the 8100 *g* membrane fraction.

The 8100 *g* membrane fraction was isolated from a mixture of 1.2×10^8 cells labeled for 3 min with [³H]uridine (28 mc/μmole, 10 μc/ml.) and 1.2×10^8 cells labeled for 120 min with [¹⁴C]-uridine (30 μc/μmole, 0.063 μc/ml.) and run to equilibrium on a 30 to 48% sucrose gradient in T buffer, as described in Materials and Methods (d). After O.D.$_{260}$ measurement, portions of each fraction were utilized for determination of radioactivity and for cytochrome oxidase assay (Materials and Methods (f)).

........, O.D.$_{260}$; --○--○--, 3 min [³H]RNA cts/min; —●——●—, 120 min [¹⁴C]RNA cts/min; —·——·—, cytochrome oxidase activity.

relative proportion to free ribosomes–polysomes, and since even a moderate enrichment of the membrane components in elements of rough endoplasmic reticulum would have been achieved only by introducing great losses of these structures and by selecting a particular fraction, it was considered necessary to use for this study the total components of the main band of the buoyant-density pattern in a sucrose gradient. This approach seemed to be justified since the other known potential source of ribosomes, mitochondria, was expected to contribute only a relatively small percentage of the total ribosomes of the membrane fraction, in view of the low amount of mitochondrial RNA (Truman & Korner, 1962; Elaev, 1966: Kroon, 1966; O'Brien & Kalf, 1967a,b). This expectation has been verified by appropriate controls (Results (e) and (f)).

(c) *Distribution of rRNA among different cytoplasmic fractions*

Table 1 shows the distribution of rRNA between the post-membranous (15,800 *g*) cytoplasmic supernatant (which contains free polysomes, monomers and "native" ribosomal subunits) and the 8100 *g* membrane fraction. (The 15,800 *g* pellet obtained from the 8100 *g* supernatant fraction contains about 10% of the free polysomes and 10 to 15% of the membrane-associated rRNA but was not systematically analyzed.) A considerable variability in the yield of rRNA associated with the two fractions and in their relative proportions was observed (Table 1): this was mainly due to

TABLE 1

Yield of ribosomal RNA and ribosomes from different cytoplasmic fractions of HeLa cells

Fraction	Ribosomal RNA	Subunits released by EDTA	Polysomes (120–350 s)
		$(\text{O.D.}_{260}\text{ units}/10^8\text{ cells})$	
Postmembranous fraction	22·6[a] (17·3–39·0)	23·3	16[d]
8100 g membrane fraction	2·9[b] (1·4–4·0)	3·3[c] (2·9–3·7)	2·2[a] (1·6–2·7) [67%[e] (64–71%)]

$$\frac{8100\ g\ \text{membrane fraction rRNA}}{\text{total rRNA}^f} \times 100 = 11\cdot5\ (6\cdot3\text{–}18\cdot1)^a\%$$

RNA was extracted by sodium dodecyl sulfate from the 15,800 g supernatant fraction and from the mitochondria–endoplasmic reticulum components of the 8100 g membrane fraction (banded in a sucrose gradient) and analyzed in a sucrose gradient in sodium dodecyl sulfate buffer. Ribosomal subunits were released from the free polysome–monomer fraction and from the 8100 g membrane fraction by EDTA treatment and isolated in a sucrose gradient in TKV. Membrane-associated polysomes were released by treatment of the 8100 g membrane fraction with 1% NaDOC and separated by centrifugation in a sucrose gradient in TKM; RNA was extracted by sodium dodecyl sulfate from the pooled fractions of the polysome region (120 to 350 s) and of the lighter components (<120 s) and run on sucrose gradients in sodium dodecyl sulfate buffer for the determination (by O.D.$_{260}$) of the proportion of rRNA in the membrane fraction which pertains to polysomes. For experimental details, see Materials and Methods (d) and (e).

[a] Average and range of 5 experiments, [b] 8 experiments and [c] 2 experiments.

[d] This amount was calculated from the average rRNA yield in the postmembranous fraction, assuming that 70% of it is in polysomes (see Results (c)).

[e] Average and range of the proportion of rRNA in the membrane fraction which pertains to polysomes.

[f] This represents the sum of the rRNA present in the 8100 g membrane fraction and in the 15,800 g supernatant; the small amount of free polysomes and membrane components sedimenting at 15,800 g would not affect significantly the ratio considered here (see text).

differences in the effectiveness of homogenization, which resulted in a varying degree of fragmentation of the cytoplasm with ensuing variable losses in the low-speed centrifugations. These losses would tend to affect preferentially the membranous structures (Blobel & Potter, 1967a); thus, although in the present work the minimal centrifugal force and time of centrifugation required to sediment nuclei were used, it is likely that the relative amount of membrane-associated rRNA was somewhat under-estimated. It appears from Table 1 that from 10 to 15% (and possibly as much as 20%) of the cytoplasmic rRNA is associated with membrane components. As is also shown in Table 1 and as will be discussed further below, the great majority of rRNA of the 8100 g membrane fraction can be accounted for by the ribosomal particles isolated from this fraction. Therefore, the average proportion of the rRNA recovered from the 15,800 g supernatant fraction and from the 8100 g membrane fraction should reflect the approximate distribution of free and membrane-associated ribosomes, respectively, in the total cytoplasmic fraction and, keeping in mind the above-mentioned cautions concerning the recovery, also in the living cell. From 65 to 70% of the ribosomes of the 8100 g membrane fraction can be recovered as polysomes after NaDOC treatment (see Results (d)). Of the ribosomes present in the 15,800 g supernatant fraction, approximately 70% are in the form of polysomes,

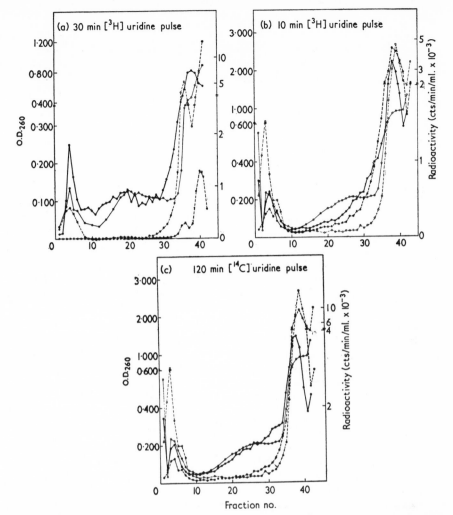

FIG. 2. RNase and EDTA sensitivity of polysomes released from the 8100 g membrane fraction by NaDOC.

(a) The 8100 g membrane fraction was isolated from 6×10^7 cells labeled for 30 min with [³H] uridine (25·4 mc/μmole, 1·25 μc/ml.), treated with 1% NaDOC, and divided into two equal parts: one-half was treated with 1 μg pancreatic RNase/ml. at 2 to 3°C for 15 min; the other half was used as a control. Each sample was run on a 15 to 30% sucrose gradient in TKM (over 3 ml. 64% sucrose) for 90 min at 24,000 rev./min. The superimposed O.D.260 and radioactivity patterns of the two gradients are shown.

(b) and (c) The 8100 g membrane fraction was isolated from a mixture of 9×10^7 cells labeled for 10 min with [³H]uridine (27·1 mc/μmole, 6·3 μc/ml.) and 9×10^7 cells labeled for 120 min with [¹⁴C]uridine (51 μc/μmole, 0·025 μc/ml.), treated with 1% NaDOC and divided into two equal parts: one-half was treated with 10^{-3} M-EDTA; the other half was used as a control. Each sample was run on a 15 to 30% sucrose gradient in low ionic strength TKV (EDTA–treated sample) or in TKM (control) under the conditions indicated in (a). The superimposed O.D.260 and ³H-radioactivity patterns of the two gradients are shown in (b); the superimposed O.D.260 and ¹⁴C-radioactivity patterns, in (c). RNA was extracted by sodium dodecyl sulfate from the pooled fractions 11 to 33 and 34 to 43 of the control gradient and run on sucrose gradients in sodium dodecyl sulfate buffer for the determination (by O.D.260) of the proportion of rRNA in the membrane fraction pertaining to polysomes (see Table 1).

(a) —○——○—, O.D.260 and ⨪●——●—, [³H]RNA cts/min of control; --○--○--, O.D.260 and --●--●--, [³H]RNA cts/min after RNase treatment.

(b) and (c) —○——○—, O.D.260 and —●——●—, [³H]RNA cts/min (b), or [¹⁴C]RNA cts/min (c) of control; --○--○--, O.D.260 and --●--●--, [³H]RNA cts/min (b) or [¹⁴C]RNA cts/min (c) after EDTA treatment.

113

about 10% in the form of 74 s† monomers, and the rest in the form of 60 and 45 s ribosomal subunits in approximately equal number (the ratio of radioactivity in the two subunits after 24-hour labeling with [³H]uridine was found to be 2).

(d) *Release of membrane-associated polysomes by treatment with sodium deoxycholate*

If the 8100 *g* membrane fraction is centrifuged through a 15 to 30% sucrose gradient in TKM, all material exhibiting o.d.$_{260}$ and radioactivity sediments rapidly to the bottom of the tube; however, it can be prevented from pelleting if a 64% sucrose cushion is used (Attardi & Attardi, 1967). Treatment of the membrane fraction with 1% NaDOC releases from these fast-sedimenting components almost all material contributing to the o.d.$_{260}$: a substantial amount of this now sediments in the form of a broad band in the region of the gradient corresponding to polysomes (120 to 350 s) and shows the characteristic RNase and EDTA sensitivity of these structures (Fig. 2); it is therefore reasonable to interpret this o.d.$_{260}$ band as consisting mainly of polysomes which have been liberated from their association with membrane components by the detergent. The small residue of fast-sedimenting material (blocked by the dense sucrose cushion) after NaDOC treatment is presumably represented mostly by structures resistant to the detergent, since the ratio of NaDOC to protein was more than adequate; a minor part of this material may be contributed by the advancing edge of the polysome band. NaDOC treatment also releases from the fast-sedimenting material most of the radioactivity incorporated into RNA: for increasing pulse length a steadily increasing proportion of the labeled membrane-associated RNA sediments after detergent action in the region of polysomes (from about 24% after a 5-min [³H]uridine pulse to about 42% after a 2-hr [¹⁴C]uridine pulse (Fig. 2(c)), with the radioactivity profile following progressivly more closely the o.d.$_{260}$ profile. The labeled components sedimenting in the polysome region show the sensitivity to RNase expected for polysomal structures (Fig. 2(a)); on the contrary, only a part of the label is associated with EDTA-sensitive structures (Fig. 2(b) and (c)). (Similar results were obtained by using EDTA concentrations from 10^{-3} to 10^{-2}.) The proportion of the label in the polysome region which is in structures sensitive to EDTA increases with pulse length: thus, it is about 32% after a 10-minute pulse with [³H]uridine (Fig. 2(b)) and becomes more than 70% after a two-hour pulse with [¹⁴C]uridine (Fig. 2(c)). The criterion of sensitivity to EDTA has been recently shown to distinguish polysomes from cosedimenting non-polysomal structures (Penman *et al.*, 1968). A reasonable interpretation of the above discussed results is that, after NaDOC treatment, in addition to polysomes, other structures derived from membrane lysis, which are resistant to EDTA and contain RNA in a form susceptible to RNase, sediment in the polysome region of the gradient. Since these EDTA-resistant structures become labeled after a very short [³H]uridine pulse, when all or the great majority of newly synthesized RNA in the cytoplasm is in mitochondria (Attardi & Attardi, 1968), it is likely that they derive from lysis of these organelles. From the proportion of rRNA in the polysome band (120 to 350 s) and in the lighter components (<120 s) (Table 1), it can be estimated that from 65 to 70% of the ribosomes in the 8100 *g* membrane fraction are released in the form

---

† The value of 74 s used in this work for the sedimentation coefficient of HeLa ribosomes has been directly determined (Attardi & Smith, 1962); for the native ribosomal subunits the values of 60 and 45 s, and for the EDTA-derived particles the values of 50 and 30 s estimated by Girard, Latham, Penman & Darnell (1965) have been used.

114

of polysomes by NaDOC action. The labeled components sedimenting in the upper third of the gradient after this treatment presumably consist in part of free RNA or RNA-containing complexes released by the lysis of mitochondria, in part of single ribosomes possibly associated with mRNA.

(e) *RNase sensitivity of membrane-associated ribosomes*

In order to obtain information concerning the possible contribution of the presence of intramitochondrial ribosomes to the results described in the previous sections, the sensitivity of the membrane-associated ribosomes to pancreatic RNase was tested: in fact, intramitochondrial ribosomes *in situ* should be protected from the action of nucleases if the mitochondrial membranes are intact (Rifkin *et al.*, 1967). Conditions of RNase digestion were found under which more than 85% of the mRNA and more than 50% of the rRNA of free polysomes are quickly made acid-soluble, while the rest is only slowly degraded (Fig. 3 and Table 2). It appears from Table 2 that, under these conditions of RNase digestion, the rRNA of membrane-bound ribosomes is made acid-soluble to the same extent (55 to 60% after a 10-min treatment) as rRNA of free polysomes. Under the same conditions, the major part (about 75%) of the mitochondrial heterogeneous RNA labeled in a three-minute [³H]-uridine pulse is not affected by the enzyme (Fig. 3 and Table 2). After solubilizing the mitochondrial membranes by NaDOC treatment, the intramitochondrial RNA becomes, on the contrary, accessible to RNase (more than 90% of the 3- to 30-min pulse-labeled RNA is quickly degraded to acid-soluble products even by low concentrations of the enzyme; see, for example, Fig. 2(a)). Since it can be reasonably assumed that the RNase-sensitive portion of the three-minute [³H]RNA provides a

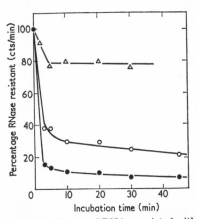

Fig. 3. Kinetics of RNase digestion *in situ* of RNA associated with mitochondria–endoplasmic reticulum components and of free polysome RNA from cells exposed for different times to [³H]uridine.

Free polysomes were isolated by sucrose gradient centrifugation of the 15,800 *g* supernatant fraction from cells labeled with [³H]uridine for 25 min (—●———●—) or 120 min (—○———○—); mitochondria–endoplasmic reticulum components were isolated by buoyant density centrifugation of the 8100 *g* membrane fraction from cells labeled with [³H]uridine for 3 min (—△———△—). 50-μl. samples (each containing less than 4 μg RNA) from the peak fraction of the free polysome profile and of the cytochrome oxidase distribution, respectively, were diluted with 1·0 ml. 0·1 M-NaCl, 0·01 M-sodium citrate and treated for different times with 50 μg pancreatic RNase/ml. at 2°C. On the axis of ordinates the percentage of cts/min which remains acid-insoluble after this treatment is indicated. RNA was extracted by sodium dodecyl sulfate from the free polysomes and from the membrane components and analyzed for proportion of radioactivity in rRNA (Table 2).

TABLE 2

Ribonuclease sensitivity in situ *of RNA in different cytoplasmic fractions from HeLa cells*

Fraction	Labeling time (min)	% of total radioactivity in rRNA[a] (A)	% of total radioactivity made acid-soluble by RNase (B)	Estimated % of radioactivity in rRNA made acid-soluble by RNase[b] (C)
Free polysomes	25 (2[c])	n.d.	87·9	—
	120 (2)	62	69·4	58
	1440 (1)	≈ 90[d]	57·9	55
8100 *g* membrane fraction	3 (8)	n.d.	26·0 (28·7[f])	—
	1560 (2)	≈ 75[e]	57·2 (55·4[f])	59 (57[f]) [47–68[g]]

Free polysomes were isolated by centrifugation in sucrose gradient of the 15,800 *g* supernatant, and the mitochondria–endoplasmic reticulum components by buoyant density centrifugation of the 8100 *g* membrane fraction (see Materials and Methods (d)) from samples of 8×10^7 to $1 \cdot 9 \times 10^8$ HeLa cells labeled for different times with [^{3}H]uridine; in one experiment, a mixture of $1 \cdot 5 \times 10^8$ cells labeled for 3 min with [^{3}H]uridine (25·5 mc/μmole, 10 μc/ml.) and 5×10^7 cells labeled for 26 hr with [^{14}C]uridine (50 μc/μmole, 0·005 μc/ml.) was used. 50- to 100-μl. samples of fractions of the free polysome distribution and of the mitochondria–endoplasmic reticulum band (each containing <6 μg RNA) were diluted with 1·0 ml. 0·1 M-NaCl, 0·01 M-sodium citrate and treated for 10 min with 50 μg pancreatic RNase/ml. at 2°C. In the double labeling experiment, the RNase-resistant radioactive material was corrected for a low level of DNA labeling by [^{14}C]-uridine by subtracting the acid-insoluble radioactive material resistant to hydrolysis by 0·5 N-NaOH for 22 hr at 30°C. RNA was released from the free polysomes and from the membrane components by sodium dodecyl sulfate and analyzed in sucrose gradient in sodium dodecyl sulfate buffer. The proportion of radioactive material associated with the two major rRNA components was estimated as described by Girard *et al.* (1965).

n.d., Not detectable.

[a] This figure does not include 5 s RNA.

[b] This estimate was made by assuming that the non-ribosomal portion of RNA from free polysomes is digested by RNase to the same extent as the 25-min labeled RNA ($\sim 88\%$), i.e.

$$C = \frac{B - (100-A) \times 0 \cdot 88}{A} \times 100,$$ where the symbols are defined in the headings of the Table.

For the non-ribosomal portion of RNA from the 8100 *g* membrane fraction, it was tentatively assumed that about 60% of it (as estimated from the fraction of heterogeneous RNA which was not released by EDTA, Fig. 4(b)) has the sensitivity to RNase of the 3 min-[^{3}H]uridine-labeled intra-mitochondrial RNA, and the rest is attacked by the enzyme to the extent of 88%, as is free polysome non-ribosomal RNA.

[c] The figures in parentheses indicate the number of experiments.

[d] This figure was estimated indirectly on the basis of the following available information: (1) almost 4% of free polysomal RNA is represented by tRNA (2 molecules per ribosome, Warner & Rich, 1964) and 5 s RNA (1 molecule per ribosome, Rosset, Monier & Julien, 1964); (2) 3 to 4% of free polysomal RNA is represented by mRNA (Soeiro, Vaughan, Warner & Darnell, 1968); and (3) the specific activity of mRNA after 24 hr growth in the presence of labeled RNA precursors is close to twice that of rRNA due to its turnover (Penman, Scherrer, Becker & Darnell, 1963; Attardi & Attardi, 1967). For the tRNA–5 s RNA components the above estimate was found to be very close to the amount directly determined from sucrose gradient analysis ($\sim 4 \cdot 3\%$).

[e] This proportion was estimated directly from sucrose gradient analysis (see, for example, Fig. 4(b) and (c)).

[f] Data pertaining to the double labeling experiment.

[g] These figures represent the minimum and maximum proportion of radioactivity in rRNA solubilized by RNase as estimated under the two extreme assumptions that all the membrane-associated heterogeneous RNA has the RNase sensitivity of free polysome non-ribosomal RNA or intramitochondrial RNA, respectively.

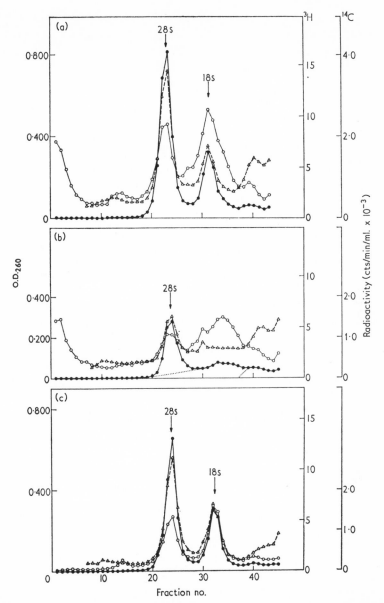

FIG. 4. Release of RNA from the 8100 *g* membrane fraction by EDTA treatment.

The 8100 *g* membrane fraction was isolated from a mixture of $1·8 \times 10^8$ cells labeled for 75 min with [^{14}C]uridine (50 μc/μmole, 0·07 μc/ml.) and 7×10^7 cells labeled for 24 hr with [^{3}H]uridine (27·1 mc/μmole, 0·3 μc/ml.) and divided into two equal parts: one-half was used as a control; the other half was treated with 90 μmoles of EDTA, pelleted, and resuspended as described in Materials and Methods (d). Each sample was run on a 30 to 48% sucrose gradient. RNA was extracted by the sodium dodecyl sulfate method from the mitochondria–endoplasmic reticulum band of the control (a) and of the EDTA-treated sample (b) and from the material released into the EDTA supernatant fraction (c), and run on a 15 to 30% sucrose gradient in sodium dodecyl sulfate buffer, as detailed in Materials and Methods (e). In (b) the dotted line was used to estimate the amount of radioactivity associated with heterogeneous RNA.

--△--△--, o.d.$_{260}$; —O——O—, 75-min [^{14}C]RNA cts/min; —●——●—, 24-hr [^{3}H]RNA cts/min.

117

maximum estimate of the fraction of mitochondria which are accessible to the enzyme and that the ribosomes contained in these mitochondria would be as susceptible to RNase as free ribosomes, these results indicate that the great majority of the ribosomes of the membrane fraction are extramitochondrial, i.e. presumably bound to the endoplasmic reticulum. This conclusion was corroborated by the results of the analysis of the effects on these ribosomes of EDTA treatment, as described below.

(f) *Response of membrane-associated ribosomes to EDTA treatment*

Investigations carried out on rough microsomes from rat liver (Sabatini *et al.*, 1966) have shown that, if these are exposed to a concentration of EDTA sufficiently high (20 μmoles per 0·5 g tissue equivalent of microsomes), essentially all the small ribosomal subunits of the membrane-bound ribosomes are released; the detachment of the large subunits requires higher concentrations of EDTA and reaches a limiting value of 50 to 60% of the original 50 s content (as can be estimated from the reported proportion of the total membrane-bound rRNA released (65 to 70%), assuming 2·6 as the weight ratio of the two subunits (Amaldi & Attardi, 1968)) upon treatment with 100 μmoles or higher amounts of EDTA. A situation similar to that described for the ribosomes of rough endoplasmic reticulum from liver, as concerns response to EDTA treatment, appears to hold for the bulk of ribosomes of the membrane fraction from HeLa cells. In fact, as shown in Figure 4 and Table 3, the analysis of RNA extracted from the EDTA-treated membrane fraction reveals that about 30% of both the pre-existing and newly synthesized (i.e. labeled after a 75-min [^{14}C]-uridine pulse) 28 s RNA originally present in the membrane components remains associated with them after exposure to the chelating agent (from 35 to 75 μmoles per gram cell equivalent of membrane fraction). On the contrary, an examination of the O.D.$_{260}$ and of the 75-minute and 24-hour labeling profiles in the sedimentation pattern of RNA from the EDTA-treated membrane fraction fails to show a clear 18 s RNA component (Fig. 4(b)). The lower ratio of radioactivity in 28 s RNA to that in 18 s RNA after 75 minutes as compared to 24-hour labeling (Fig. 4(a)) is due to the asynchrony of arrival in the cytoplasm from the nucleus of the two newly synthesized rRNA components (Girard *et al.*, 1965). The broad band of heterogeneous

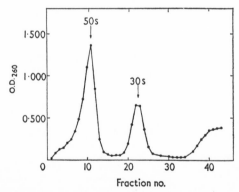

FIG. 5. Sedimentation pattern of ribosomal subunits released from the 8100 *g* membrane fraction by EDTA treatment. The 8100 *g* membrane fraction was isolated from 2·6 × 10⁸ cells; after treatment with 100 μmoles of EDTA, the membrane components were spun down and the supernatant fraction was run on a 15 to 30% sucrose gradient in TKV, as described in Materials and Methods (d).

TABLE 3

Release of RNA from the 8100 g membrane fraction by EDTA treatment

Experiment	Precursor	Labeling time (min)	Fraction (after EDTA treatment)	Ribosomal RNA (cts/min)			Heterogeneous RNA[a]
				Total	28 s	18 s	
1	[³H]Uridine	3	pellet	n.d.	n.d.	n.d.	930 (89%)
			supernatant	n.d.	n.d.	n.d.	110 (11%)
	[¹⁴C]Uridine	75	pellet	660	660 (31%)	n.d.	8130 (80%)
			supernatant	3340	1440 (69%)	1900	2000 (20%)
2	[¹⁴C]Uridine	75	pellet	630	630 (28%)	n.d.	11,000 (83%)
			supernatant	3400	1620 (72%)	1780	2280 (17%)
	[³H]Uridine	1440	pellet	6550	6550 (31%)	n.d.	7000
			supernatant	21,950	14,250 (69%)	7700	≈1600

In experiment 1, the 8100 g membrane fraction was isolated from a mixture of 7×10^7 cells labeled for 3 min with [³H]uridine (28 mc/µmole, 10 µc/ml.) and 7×10^7 cells labeled for 75 min with [¹⁴C]uridine (30 µc/µmole, 0·063 µc/ml.). After treatment with 40 µmoles of EDTA, the membrane fraction was pelleted, resuspended, and run on a 30 to 48% sucrose gradient, as described in Materials and Methods. RNA was extracted by sodium dodecyl sulfate from the membrane components of the main band in the gradient and from the material released into the EDTA supernatant fraction, and run on 15 to 30% sucrose gradients in sodium dodecyl sulfate buffer, as detailed in Materials and Methods (e). The distribution of radioactivity between the two major rRNA components and heterogeneous RNA was estimated as described by Girard et al. (1965). The data of experiment 2, obtained in the same way, pertain to the RNA extracted from the EDTA-treated membrane fraction used in the experiment of Fig. 4.

n.d., Not detectable.

[a] Includes all non-ribosomal RNA from 7 s to more than 50 s.

TABLE 4

Ratio of 28 to 18 s RNA and of 50 to 30 s ribosomal subunits in different cytoplasmic fractions from HeLa cells

Fraction		Ratio of 28 to 18 s RNA[a]	Ratio of 50 to 30 s subunits[a]
8100 *g* Membrane fraction	Total	2·65[c]	2·62[c]
	Material released by EDTA	1·85	1·83
Free polysomes		2·47	2·75
Total ribosome-polysome fraction		2·43[d]	2·65[d]
Expected from molecular weight data[b]		2·5 to 2·6	

The RNA was extracted from the mitochondria–endoplasmic reticulum components of the EDTA-treated membrane fraction (banded in sucrose gradient), from the material released by EDTA and from free polysomes by sodium dodecyl sulfate, as detailed in Materials and Methods (e). The ribosomal subunits were isolated as described in Materials and Methods (d). The ratio of 28 to 18 s RNA in free polysomes and that of 50 to 30 s subunits in the EDTA supernatant fraction and in free polysomes were determined on the basis of optical density at 260 mμ after correction for the different base composition of the two rRNA components (Amaldi & Attardi, 1968). The ratio of 28 to 18 s RNA in the material released by EDTA was determined on the basis of distribution of radioactivity after 24-hr [^{3}H]uridine labeling (no correction for the different base composition of the two rRNA species was needed because the sum of U + C + ψ is almost identical in the two components, and C and U (and presumably ψ) have about the same specific activity after long exposure of the cells to labeled uridine (Salzman & Sebring, 1959). The slightly lower values obtained for the ratio of the two rRNA components, as compared to the ratio of the two ribosomal subunits, in free polysomes and in the total ribosome–polysome fraction may indicate a small conversion (less than 3%) of 28 s RNA to molecules sedimenting as 18 s RNA.

[a] Defined in the text.

[b] The figures given here represent the ratio of 28 to 18 s (of 50 to 30 s subunits) expected for equimolar amounts of the two rRNA species from their molecular weight ratio (Amaldi & Attardi, 1968).

[c] These figures represent the ratios determined on the material released by EDTA after correction for the amount of 28 s RNA (50 s subunits) remaining in the EDTA-treated membrane fraction (on the average, about 30% of total membrane-associated 28 s RNA (50 s subunits) (see Table 3)).

[d] Average values determined on the basis of O.D.$_{260}$ ratios and ^{32}P-labeling ratios (Amaldi & Attardi, 1968).

RNA sedimenting in the region 9 to 23 s and the heavier polydisperse components (up to more than 50 s) represent, in part at least, mitochondrial RNA (see Discussion).

Figure 5 shows the sedimentation pattern of the material released from the membrane fraction of HeLa cells by EDTA treatment: one recognizes the two ribosomal subunits and a small amount of slower sedimenting ultraviolet-absorbing material near the meniscus. The ratio of major to minor subunits (defined as the ratio of total number of nucleotides contained in the two classes) released by EDTA from the 8100 *g* membrane fraction is about 1·83 (Table 4). This ratio is considerably lower than that of 2·75 found for the subunits derived from free polysomes treated under the same conditions, and that of 2·65 previously reported for the subunits derived from the total ribosome–polysome fraction (Amaldi & Attardi, 1968): this difference can be accounted for completely by the amount of 28 s RNA remaining in the EDTA-treated membrane fraction, a finding which strongly suggests that this residual 28 s RNA is in the form of 50 s subunits. In the same way, the figure of 1·85

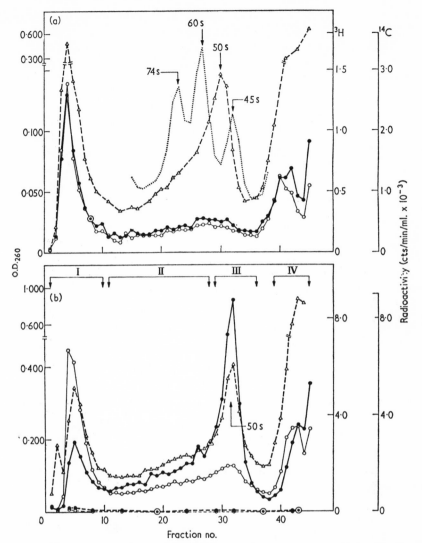

Fig. 6. Sedimentation pattern of the NaDOC lysate of EDTA-treated membrane fraction from cells exposed for different times to labeled uridine.

The 8100 g membrane fraction was isolated, in the experiment shown in (a), from a mixture of 3×10^7 cells labeled for 5 min with [³H]uridine (27·1 mc/μmole, 6·3 μc/ml.) and 3×10^7 cells labeled for 30 min with [¹⁴C]uridine (51 μc/μmole, 0·063 μc/ml.), and in the experiment shown in (b), from a mixture of $1·0 \times 10^8$ cells labeled for 75 min with [¹⁴C]uridine (50 μc/mole, 0·07 μc/ml.) and 4×10^7 cells labeled for 24 hr with [³H]uridine (27·1 mc/μmole, 0·3 μc/ml.). After exposure to EDTA, the membrane fraction was pelleted, resuspended in 0·25 M-sucrose in T buffer (2·0 ml. in Expt. (a) and 1·0 ml. in Expt. (b)), treated with 1% NaDOC, and run on a 15 to 30% sucrose gradient (over 3 ml. 64% sucrose) in TKV, as detailed in Materials and Methods (d). In (a) a sample of the 15,800 g supernatant fraction was run at the same time on a 15 to 30% sucrose gradient in TKM to provide the o.d.₂₆₀ profile of free monomers and native ribosomal subunits.

For the determination of incorporation of labeled uridine into alkali-stable material, portions of

121

found for the ratio of 28 to 18 s RNA (defined as above) in the material released by EDTA, after correction for the residual membrane-associated 28 s RNA, becomes very close to the ratio of the two major rRNA species in free polysomes (Table 4). No obvious loss of intramitochondrial RNA components appears to occur as a result of EDTA treatment. Thus, under conditions which release from the membrane fraction almost all the 30 s ribosomal subunits, about 90% of the heterogeneous RNA labeled after a three-minute [^{3}H]uridine pulse (when 70% or more of the label is intramitochondrial (Attardi & Attardi, 1968)) and about 80% of the 75-minute ^{14}C-labeled heterogeneous RNA (of which at least 60% is intramitochondrial) remain in the EDTA-treated membrane fraction (Table 3). Although exposure of the membrane components to EDTA results in a slight decrease ($\sim$ 0·01 g/ml.) in density in sucrose gradient of the mitochondria–rough endoplasmic reticulum band (as revealed by the cytochrome oxidase, o.d.$_{260}$ and radioactivity profiles), no appreciable effect on the susceptibility *in situ* to RNase of the membrane-associated 75-minute [^{14}C]uridine pulse-labeled heterogeneous RNA was detected after this treatment: this finding speaks against the possibility of damage to mitochondrial membranes resulting in loss of intramitochondrial ribosomes.

The results analyzed above of the effects of EDTA treatment on membrane-associated ribosomes confirm the conclusion of the RNase-sensitivity experiments, indicating that the great majority of these ribosomes are bound to the endoplasmic reticulum. The amount of pre-existing heterogeneous RNA not removed by EDTA, estimated on the basis of the distribution of the 24-hour labeled material (Fig. 4(b) and (c)) and assuming that the specific activity of the heterogeneous RNA after this labeling time is about double that of rRNA (due to its turnover (Attardi & Attardi, 1967)), represents 9 to 10% of the total membrane-associated RNA. As mentioned above, a part of this RNA, and possibly the majority, is intramitochondrial (see Discussion).

The nature of the relatively minor portion of labeled heterogeneous RNA of the membrane fraction which is found in the supernatant fraction after EDTA treatment (about 20% after 75-min labeling), remains to be established; it seems likely that it contains mRNA of membrane-bound polysomes which has been released together with the ribosomal subunits.

(g) *Analysis of the material released by sodium deoxycholate from EDTA-treated membrane fraction*

In order to try to isolate the 50 s subunits which remain associated with membrane components after EDTA treatment (see preceding section), the 8100 *g* membrane

---

individual fractions of (b) were treated with 0·5 N-NaOH for 22 hr at 30°C, neutralized with HCl, and then precipitated with 5% trichloroacetic acid.

In (b), the components corresponding to the portions of the sucrose gradient pattern indicated by arrows were utilized for RNA analysis (see Fig. 7).

(a) --△--△--, o.d.$_{260}$; —○——○—, 30-min [^{14}C]RNA cts/min; —●——●—, 5-min [^{3}H]RNA cts/min.

(b) --△--△--, o.d.$_{260}$; —○——○—, total 75-min [^{14}C]RNA cts/min; —●——●—, total 24-hr [^{3}H]RNA cts/min; --○--○--, alkali-resistant ^{14}C cts/min; --●--●-- alkali-resistant ^{3}H cts/min.

5

fraction from cells labeled for various times with [³H]- or [¹⁴C]uridine was treated with 3×10^{-2} M-EDTA, washed with 1.5×10^{-2} M-EDTA, lysed with 1% NaDOC, and run on a sucrose gradient. The O.D.$_{260}$ profile thus obtained (Fig. 6) showed a well-defined peak with the approximate sedimentation coefficient of 50 s (estimated by using the native ribosomal subunits 60 and 45 s present in the 15,800 **g** supernatant fraction as reference markers) and faster sedimenting components extending down to the region of 130 s and more (the heaviest material being prevented from pelleting by the dense sucrose cushion at the bottom of the tube). The radioactivity incorporated into RNA in a short pulse (which is mostly intramitochondrial) appeared to be distributed rather uniformly in the region of the gradient from 30 to about 130 s (Fig. 6(a)); there was a somewhat greater accumulation of label in the components sedimenting slower than 30 s and, furthermore, a considerable proportion of radioactive material sedimented with the material heavier than 130 s (presumably these fast sedimenting structures correspond to the rapidly labeled material released by NaDOC lysis of mitochondria which still sediments in the polysome region after EDTA treatment (Fig. 2(b) and (c)). No labeling of the 50 s peak was observed after exposing the cells to [¹⁴C]uridine for up to 30 minutes; after a 75-minute pulse, some radioactivity appeared to be associated with the 50 s component (Fig. 6(b)). The labeling of the latter became progressively more pronounced with increasing time of exposure of the cells to the radioactive precursor (Fig. 6(b)). All the labeled components in the gradient appeared to be alkali-sensitive. It seemed likely that the 50 s peak consisted of large ribosomal subunits from membrane-bound ribosomes, which had not been detached by EDTA treatment. Consistent with this possibility was the delay in appearance of label in this peak, which agreed with the time of arrival of newly synthesized 28 s RNA to the endoplasmic reticulum (Attardi & Attardi, 1967). An analysis of the sedimentation pattern (O.D.$_{260}$ and 24-hr labeling profiles) of the RNA extracted from the 50 s peak region of the gradient showed indeed a well-defined 28 s RNA component and, in addition, heterogeneous material sedimenting in the region 7 to 28 s (Fig. 7(c)). A surprising finding, however, was that a very similar pattern, with a 28 s RNA peak and slower sedimenting heterogeneous material (Fig. 7(b)), was exhibited by the RNA extracted from the structures sedimenting between 65 and 130 s in the sucrose-gradient centrifugation analysis shown in Figure 6(b), and also by the RNA derived from the heaviest components (>130 s) (in the latter case some heterogeneous RNA with sedimentation coefficients from 30 to more than 50 s was also present) (Fig. 7(a)). The RNA extracted from the NaDOC-released components with $S < 20$ s showed, on the contrary, a high peak sedimenting at about 6 s, with small amounts of heavier heterogeneous material and of 28 s RNA (Fig. 7(d)).

An analysis of the distribution of 28 s RNA among the various components liberated by NaDOC from the EDTA-treated membrane fraction (Fig. 7) revealed that only about 37% of the large rRNA species remaining in this fraction after exposure to EDTA could be recovered from the 50 s peak region (Fig. 6(b)), the remainder being mostly in structures sedimenting faster than 50 s (up to more than 130 s). Increasing the relative amount of NaDOC from 10 mg (Fig. 6(b)) to 45 mg (Fig. 6(a)) per gram cell equivalent of membrane fraction did not promote the release of free 50 s particles as judged from the relative amount of O.D.$_{260}$ in the 50 s peak. As in the case of the RNA extracted from the total membrane fraction after exposure to EDTA (Fig. 4(b)), an analysis of the O.D.$_{260}$ and of the 24-hour labeling profiles

123

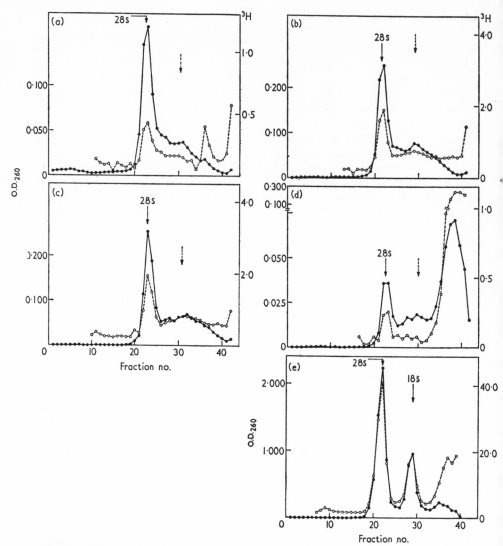

Fig. 7. Sedimentation pattern of RNA extracted from the various components released by NaDOC from the EDTA-treated membrane fraction. (Only the O.D.$_{260}$ and the 24-hr labeling profiles are shown here for the sake of simplicity.)

RNA was released by sodium dodecyl sulfate from the components of the NaDOC lysate of the EDTA-treated membrane fraction shown in Fig. 6(b), and run on sucrose gradients in sodium dodecyl sulfate buffer, as specified in Materials and Methods (e): the patterns (a) to (d) correspond respectively, to sections I to IV in the gradient of Fig. 6(b). The dashed arrow indicates the presumptive position of 18 s RNA. In (e) the sedimentation pattern of free polysomal RNA extracted from the same cells and run on a sucrose gradient under the same conditions is shown to provide a position marker for the two major rRNA species.

--○--○--, O.D.$_{260}$; —●——●—, 24-hr [^{3}H]RNA cts/min.

124

in the 14 to 20 s region of the sedimentation patterns of RNA from the various NaDOC lysis products of the EDTA-treated membrane fraction (Fig. 7) did not show a clear 18 s RNA component.

4. Discussion

The electron microscopic examination of sections of HeLa cells has confirmed the observations made by others (Epstein, 1961; Journey & Goldstein, 1961; Fuse et al., 1963) concerning the presence in these cells of tubules, vesicles and cisternae of rough endoplasmic reticulum. In HeLa cells, as in other rapidly multiplying cells, the membrane-bound ribosomes represent a relatively minor part of total ribosomes: a rough visual estimate of their proportion in the electron microscope pictures would be between 10 and 20%.

In the present work, most of the elements of rough endoplasmic reticulum were recovered, together with the bulk of mitochondria, in the 8100 g membrane fraction. Attempts to resolve rough endoplasmic reticulum from mitochondria either by differential centrifugation or by sedimentation-velocity or buoyant-density fractionation in sucrose gradient gave unsatisfactory results. It was found that even a partial separation of the endoplasmic reticulum involved drastic losses of these structures, in addition to introducing a disturbing selection factor. Since mitochondria are known to contain very small amounts of RNA (Truman & Korner, 1962; Elaev, 1966, Kroon, 1966; O'Brien & Kalf, 1967a,b) and were therefore expected to contribute a relatively small proportion of the population of membrane-associated ribosomes, it was considered justifiable for the purposes of the present work to utilize the components of the 8100 g membrane fraction with a buoyant density in sucrose gradient of 1·16 to 1·20 g/ml. (which include both rough endoplasmic reticulum and mitochondria).

The results obtained in this work on the response of the membrane-associated ribosomes to RNase and EDTA treatment have confirmed the validity of this approach. Thus, under conditions of pancreatic RNase digestion which left the intramitochondrial heterogeneous RNA mostly intact, the RNA of the membrane-associated ribosomes appeared to be degraded to acid-soluble products to approximately the same extent (55 to 60%) as rRNA of free polysomes. Furthermore, exposure to EDTA released from the membrane components into the supernatant fraction almost all the small ribosomal subunits. Under the same conditions, the intramitochondrial heterogeneous RNA was not removed, and judging from its sensitivity to externally added RNase, there was no apparent damage to the mitochondrial membranes which could have resulted in loss of intramitochondrial ribosomal subunits. On the other hand, the amount of residual 28 s RNA in the EDTA-treated membrane fraction was that expected for membrane-stuck 50 s subunits from ribosomes of rough endoplasmic reticulum (see below). These results, therefore, point to the association with endoplasmic reticulum of the bulk of the ribosomes of the membrane fraction. The analysis carried out in this work was not sensitive enough to detect a relatively small amount of intramitochondrial ribosomes. Experiments carried out with purified mitochondria will be necessary to investigate the existence of intramitochondrial ribosomes, which is suggested by the analogy with lower eukaryotic cells (Kuntzel & Noll, 1967; Rifkin et al., 1967; Dure et al., 1967; Rogers et al., 1967; Wintersberger, 1967; Clark-Walker & Linnane, 1967), by the protein synthetic capacity of mitochondria in vitro (Roodyn et al., 1961,1962; Craddock & Simpson, 1961; Truman & Korner, 1962; Kroon,

125

1963*a,b,c*; Wheeldon & Lehninger, 1966) and by the apparent presence in rat liver mitochondria of specific tRNA species (Buck & Nass, 1968; Smith & Marcker, 1968).

On the basis of the amount of rRNA extracted from the 8100 g membrane fraction and from the small proportion of slower sedimenting mitochondria and endoplasmic reticulum present in the 15,800 g pellet, it could be estimated that from 10 to 15% (and possibly as many as 20%) of the total ribosomes in HeLa cells are associated with endoplasmic reticulum. Preferential losses of membrane-associated ribosomes in the low-speed centrifugations (Blobel & Potter, 1967*a*) would, however, lead to an under-estimate of their relative amount in the living cell. It should be pointed out that the concentration of Mg^{2+} during the cell homogenization and subsequent fractionation steps was kept to the minimum compatible with stability of polysomes, namely, 1.5×10^{-4} M, in order to discourage any aggregation of free ribosomes and polysomes with membrane components. The observation that the ribosomes of the membrane fraction from HeLa cells behave, as concerns response to EDTA treatment, similarly to ribosomes of the rough endoplasmic reticulum from liver and differently from free ribosomes (see below) supports the conclusion that the ribosomes isolated with this fraction are associated with membrane elements in the intact cell. 65 to 70% of these ribosomes are recovered in the form of polysomes after NaDOC treatment of the membrane fraction and are presumably engaged in protein synthesis *in vivo*: the polysomes released by the detergent show the same RNase and EDTA sensitivity as free polysomes. Unlike the situation described for rat liver (Blobel & Potter, 1966; Howell *et al.*, 1964), in HeLa cells exposure to NaDOC of the rough endoplasmic reticulum structures sedimenting with mitochondria does not lead to an extensive breakdown of membrane-bound polysomes by nuclease action. However, the possibility of a partial degradation of polysomal mRNA after treatment with NaDOC cannot be excluded: as a matter of fact, the release of lysosomal RNase by the action of detergents has been recently described (Penman, Vesco & Penman, 1968). There-fore, the amount of polysomes recovered from the membrane fraction by NaDOC treatment can only provide a minimum estimate of the fraction of membrane-bound ribosomes which are associated with mRNA in the living cell. As was mentioned above, the rRNA of endoplasmic reticulum-bound ribosomes showed the same sensitivity to RNase digestion, under the conditions used in the present work, as rRNA of free polysomes. The apparent discrepancy between this result and the previously reported relative RNase resistance of membrane-bound ribosomes from liver (Blobel & Potter, 1967*b*) is presumably due to the different conditions of digestion used here: particularly significant in this regard may be the higher ionic strength and the presence of a chelating agent in the medium (which would probably cause detachment of ribosomes from the membranes and removal of adventitious ribosomal proteins (Warner & Péne, 1966)) and the much higher (at least 15 times) RNase to RNA ratio.

As discussed earlier, treatment with EDTA (35 to 70 μmoles/g cell equivalent of membrane fraction) releases essentially all the 30 s subunits and about 70% of the 50 s subunits of the endoplasmic reticulum-bound ribosomes. The behavior of these ribosomes upon exposure to EDTA appears, therefore, to be similar to that described for ribosomes of the rough endoplasmic reticulum from rat liver (Sabatini *et al.*, 1966). It is not known which factor(s) are involved in the EDTA-resistant sticking of the 50 s subunits to the membrane. There seems to be a functional relationship between strong attachment and activity in protein synthesis of membrane-bound

126

ribosomes (Sabatini *et al.*, 1966), although the growing polypeptide chain does not seem to be itself the "hook" (Redman & Sabatini, 1966). It is possible that conformational changes induced in the ribosome, or in the membrane or in both, by the formation of the active complex cause a better fit between a specific receptor on the membrane and a binding site on the ribosome. NaDOC liberates from the EDTA-treated membrane fraction only 35 to 40% of the membrane-stuck subunits as free 50 s particles; the remainder are contained in structures sedimenting faster than 50 s (up to more than 130 s). The nature of these heavier structures is uncertain. The virtual absence of small ribosomal subunits and the resistance to EDTA argues against their being polysomes. Their presence may reflect the tendency of some of the EDTA-resistant 50 s subunits (perhaps those with more complete polypeptide chains) to stick to residues of membrane elements or to aggregate with each other, possibly by interacting nascent chains. It should be mentioned in this connection that, in the case of rough microsomes from rat liver, the particulate material released by NaDOC after EDTA treatment (which consists mainly of a 50 s component and of 80 to 90 s heterogeneous material) contains a major portion of the newly synthesized protein and/or nascent polypeptides labeled *in vivo* during a one-minute [³H]leucine pulse (Sabatini *et al.*, 1966).

As concerns the mRNA of membrane-bound polysomes, one would expect it to be released by EDTA together with the 30 s subribosomal particles, in view of the involvement of the small subunits in the binding of messenger (see review by Attardi, 1967) and in view of the evidence indicating that the ribosomes are attached to the membrane through their 50 s subunits and behave as fixed points, with the mRNA being the moving component of the active complex (Sabatini *et al.*, 1966; Blobel & Potter, 1967b). It is likely that the relatively minor portion of labeled heterogeneous RNA of the membrane fraction which is found in the supernatant fraction after EDTA treatment consists, at least in part, of this mRNA.

The observation that the intramitochondrial heterogeneous RNA labeled in a short pulse (which is homologous to mitochondrial DNA (Attardi & Attardi, 1968)) is not removed by EDTA treatment (Table 3) indicates that a portion, possibly the majority, of the heterogeneous RNA remaining in the EDTA-treated membrane fraction is represented by mitochondrial RNA. However, direct evidence obtained with purified subcellular components is needed to establish both the fate, after exposure to EDTA, of the mRNA of endoplasmic reticulum-bound polysomes and the contribution of non-mitochondrial RNA components to the EDTA-resistant fraction.

These investigations were supported by a research grant from the U.S. Public Health Service (GM-11726). We thank Mrs L. Wenzel and Mrs B. Keeley for their invaluable help throughout this work.

REFERENCES

Amaldi, F. & Attardi, G. (1968). *J. Mol. Biol.* **33**, 737.
André, J. & Marinozzi, V. (1965). *J. Microscopie*, **4**, 615.
Attardi, B. & Attardi, G. (1967). *Proc. Nat. Acad. Sci., Wash.* **58**, 1051.
Attardi, G. (1967). *Ann. Rev. Microbiol.* **21**, 383.
Attardi, G. & Attardi, B. (1968). *Proc. Nat. Acad. Sci., Wash.* **61**, 261.
Attardi, G., Parnas, H., Hwang, M-I. H. & Attardi, B. (1966). *J. Mol. Biol.* **20**, 145.
Attardi, G. & Smith, J. D. (1962). *Cold Spr. Harb. Symp. Quant. Biol.* **27**, 271.

Blobel, G. & Potter, V. R. (1966). *Proc. Nat. Acad. Sci., Wash.* **55**, 1283.
Blobel, G. & Potter, V. R. (1967a). *J. Mol. Biol.* **26**, 279.
Blobel, G. & Potter, V. R. (1967b). *J. Mol. Biol.* **26**, 293.
Buck, C. A. & Nass, M. M. K. (1968) *Proc. Nat. Acad. Sci., Wash.* **60**, 1045.
Cammarano, P., Giudice, G. & Lukes, B. (1965). *Biochem. Biophys. Res. Comm.* **19**, 487.
Clark-Walker, G. D. & Linnane, A. W. (1967). *J. Cell Biol.* **34**, 1.
Craddock, V. M. & Simpson, M. V. (1961). *Biochem. J.* **80**, 348.
Dallner, G., Siekevitz, P. & Palade, G. E. (1966). *J. Cell Biol.* **30**, 73.
Dubin, D. T. & Brown, R. E. (1967). *Biochim. biophys. Acta,* **145**, 538.
Dure, L. S., Epler, J. L. & Barnett, W. E. (1967). *Proc. Nat. Acad. Sci., Wash.* **58**, 1883.
Elaev, I. R. (1966). *Biokhimiya,* **31**, 234.
Epstein, M. A. (1961). *J. Biophys. Biochem. Cytol.* **10**, 153.
Fuse, Y., Price, Z. & Carpenter, C. M. (1963). *Cancer Res.* **23**, 1658.
Gilbert, W. (1963). *J. Mol. Biol.* **6**, 389.
Girard, M., Latham, H., Penman, S. & Darnell, J. E. (1965). *J. Mol. Biol.* **11**, 187.
Henshaw, E. C., Bojarski, T. B. & Hiatt, H. H. (1963). *J. Mol. Biol.* **7**, 122.
Howell, R. R., Loeb, J. N. & Tomkins, G. M. (1964). *Proc. Nat. Acad. Sci., Wash.* **52**, 1241.
Journey, L. J. & Goldstein, M. N. (1961). *Cancer Res.* **21**, 929.
Kroon, A. M. (1963a). *Biochim. biophys. Acta,* **69**, 184.
Kroon, A. M. (1963b). *Biochim. biophys. Acta,* **72**, 391.
Kroon, A. M. (1963c). *Biochim. biophys. Acta,* **76**, 165.
Kroon, A. M. (1966). In *Regulation of Metabolic Processes in Mitochondria,* ed. by J. M. Tager, S. Papa, E. Quagliariello & E. C. Slater, p. 397. Amsterdam: Elsevier Publishing Co.
Küntzel, H. & Noll, H. (1967). *Nature,* **215**, 1340.
Loeb, J. N., Howell, R. R. & Tomkins, G. M. (1965). *Science,* **149**, 1093.
Loeb, J. N., Howell, R. R. & Tomkins, G. M. (1967). *J. Biol. Chem.* **242**, 2069.
Manganiello, V. C. & Phillips, A. H. (1965). *J. Biol. Chem.* **240**, 3951.
Moulé, Y., Rouiller, C. & Chauveau, J. (1960). *J. Biophys. Biochem. Cytol.* **7**, 547.
O'Brien, T. W. & Kalf, G. F. (1967a) *J. Biol. Chem.* **242**, 2172.
O'Brien, T. W. & Kalf, G. F. (1967b). *J. Biol. Chem.* **242**, 2180.
Palade, G. E. (1952). *J. Exp. Med.* **95**, 285.
Palade, G. E. (1956). *J. Biophys. Biochem. Cytol.* **2**, (no. 4, suppl.), 85.
Palade, G. E. (1958). In *Microsomal Particles and Protein Synthesis,* ed. by R. B. Roberts, p. 36. New York: Pergamon Press.
Parsons, D. F., Williams, G. R. & Chance, B. (1966). *Ann. N.Y. Acad. Sci.* **137**, 643.
Penman, S., Scherrer, K., Becker, Y. & Darnell, J. E. (1963). *Proc. Nat. Acad. Sci., Wash.* **49**, 654.
Penman, S., Vesco, C. & Penman, M. (1968). *J. Mol. Biol.* **34**, 49.
Porter, K. R. (1961). In *The Cell,* ed. by J. Brachet & A. E. Mirsky, p. 621. New York: Academic Press.
Redman, C. M. & Sabatini, D. D. (1966). *Proc. Nat. Acad. Sci., Wash.* **56**, 608.
Redman, C., Siekevitz, P. & Palade, G. E. (1966). *J. Biol. Chem.* **241**, 1150.
Rendi, R. & Warner, R. C. (1960). *Ann. N.Y. Acad. Sci.* **88**, 741.
Reynolds, E. S. (1963). *J. Cell Biol.* **17**, 208.
Rifkin, M. R., Wood, D. D. & Luck, D. J. L. (1967). *Proc. Nat. Acad. Sci., Wash.* **58**, 1025.
Rogers, P. J., Preston, B. N., Titchener, E. B. & Linnane, A. W. (1967). *Biochem., Biophys. Res. Comm.* **27**, 405.
Roodyn, D. B. Reis, P. J. & Work, T. S. (1961). *Biochem. J.* **80**, 9.
Roodyn, D. B., Suttie, J. W. & Work, T. S. (1962). *Biochem. J.* **83**, 29.
Rosset, R., Monier, R. & Julien, J. (1964). *Bull. Soc. chim. Biol.* **46**, 87.
Sabatini, D. D., Tashiro, Y. & Palade, G. E. (1966). *J. Mol. Biol.* **19**, 503.
Salzman, N. P. & Sebring, E. D. (1959). *Arch. Biochem. Biophys.* **84**, 143.
Siekevitz, P. & Palade, G. E. (1960). *J. Biophys. Biochem. Cytol.* **7**, 619.
Smith, A. E. & Marcker, K. A. (1968). *J. Mol. Biol.* **38**, 241.
Smith, L. (1954). *Arch. Biochem. Biophys.* **50**, 285.
Soeiro, R., Vaughan, M. H., Warner, J. R. & Darnell, J. E. (1968). *J. Cell Biol.* **39**, 112.

128

Swift, H. (1965). *Amer. Natur.* **99**, 201.
Truman, D. E. S. (1963). *Exp. Cell Res.* **31**, 313.
Truman, D. E. S. & Korner, A. (1962). *Biochem. J.* **83**, 588.
Warner, J. R. & Péne, M. G. (1966). *Biochim. biophys. Acta,* **129**, 359.
Warner, J. R. & Rich, A. (1964). *Proc. Nat. Acad. Sci., Wash.* **51**, 1134.
Watson, M. L. & Aldridge, W. G. (1964). *J. Histochem. Cytochem.* **12**, 96.
Webb, T. E., Blobel, G. & Potter, V. R. (1964). *Cancer Res.* **24**, 1229.
Wheeldon, L. W. & Lehninger, A. L. (1966). *Biochemistry,* **5**, 3533.
Wintersberger, E. (1967). *Z. physiol. Chem.* **348**, 1701.

Relationship between HeLa Cell Ribosomal RNA

and its Precursors studied by High Resolution

RNA–DNA Hybridization

Ph. Jeanteur

G. Attardi

1. Introduction

It is well established that in animal cells the synthesis of rRNA‡ occurs in the form of high molecular weight precursor molecules with a sedimentation coefficient of about 45 s (Perry, 1962; Scherrer, Latham & Darnell, 1963; Rake & Graham, 1964; Penman, 1966; Muramatsu, Hodnett & Busch, 1966; see review by Perry, 1967). Furthermore, a considerable amount of experimental evidence supports

---

‡ Abbreviations used: rRNA, ribosomal RNA; rDNA, DNA stretches homologous to 45 s rRNA precursor; dodecyl SO₄, sodium dodecyl sulfate.

the idea that the 45 s molecules are precursors of both 28 and 18 s RNA (Scherrer et al., 1963; Perry, 1965; Penman, 1966; Greenberg & Penman, 1966; Zimmerman & Holler, 1966,1967; Wagner, Penman & Ingram, 1967). Previous base composition and partial sequence studies carried out in this laboratory have led to the conclusion that the conversion of 45 s RNA to mature rRNA is a non-conservative process (Amaldi & Attardi, 1968; Jeanteur, Amaldi & Attardi, 1968); thus, it has been shown that an "average" 45 s molecule contains *at most* the sequences of one 28 s molecule and one 18 s molecule, and that consequently about 50% of its length is discarded in the maturation process†. Similarly, about 30% of the 32 s molecule, which appears as an intermediate in the process leading to mature 28 s RNA, is represented by sequences of non-ribosomal type which are not conserved during the conversion to 28 s RNA. Observations indicating the non-conservative nature of the processing of the 45 s and 32 s rRNA precursors have been also reported by other investigators (Weinberg, Loening, Willems & Penman, 1967; Vaughan, Soeiro, Warner & Darnell, 1967; Roberts & D'Ari, 1968; Willems, Wagner, Laing & Penman, 1968; McConkey & Hopkins, 1969).

In the present work, RNA–DNA hybridization experiments have been carried out with highly purified 18, 28, 32 and 45s RNA with the aim of analyzing, by a different approach from those previously used, the relationship between rRNA and its precursors. The methodology previously described for the isolation and analysis of specificity of RNA–DNA hybrids (Attardi, Huang & Kabat, 1965a) has been used here. The results obtained are consistent with the model according to which about 35% of the 45 s molecule is represented by the 28 s polynucleotide stretch, about 13% by the 18 s stretch, and the rest by sequences of non-ribosomal type; furthermore, they support the conclusion of the base sequence studies, indicating that about 30% of the 32 s molecule cannot be accounted for by sequences of mature rRNA.

2. Materials and Methods

(a) *Cells and method of growth*

The method of growth of HeLa cells (S3 clonal strain) in suspension has been described previously (Amaldi & Attardi, 1968).

(b) *Buffers*

The buffer designations are: (1) Dodecyl SO_4 buffer (Gilbert, 1963): 0·01 M-Tris buffer (pH 7·0), 0·1 M-NaCl, 0·001 M-EDTA, 0·5% dodecyl SO_4; (2) SSC (standard saline citrate): 0·15 M-NaCl, 0·015 M-sodium citrate; (3) dissociation medium: 0·01 M-potassium phosphate buffer (pH 8·0), 2% formaldehyde, 0·1% dodecyl SO_4; (4) PO_4–NaCl–formaldehyde buffer: 0·02 M-potassium phosphate buffer (pH 7·4), 0·1 M-NaCl, 1% formaldehyde; (5) acetate–NaCl buffer: 0·01 M-acetate buffer (pH 5·0), 0·1 M-NaCl; (6) Tris–K–Mg: 0·05 M-Tris buffer (pH 6·7), 0·025 M-KCl, 0·0025 M-$MgCl_2$.

† In the present work the molecular weights of 18, 28, 32 and 45 s RNA have been considered to be about 0·6, 1·6, 2·2 and 4·6 × 10⁶, respectively (Weinberg *et al.*, 1967; Jeanteur *et al.*, 1968).

(c) Labeling conditions

Labeling with ^{32}P of the rRNA precursors was carried out by exposing cells for 4 hr to [^{32}P]orthophosphate in medium containing 5×10^{-7} to 10^{-5} M-phosphate. Labeling of rRNA to a high specific activity with [^{32}P]orthophosphate and chase with unlabeled phosphate to reduce the specific activity of the labile RNA species were done as previously reported (Attardi, Huang & Kabat, 1965b). For long-term labeling of rRNA with [^{3}H]-uridine, HeLa cells (5×10^4/ml.) were incubated for 48 hr in the presence of 2·5 µc [5-^{3}H]uridine/ml. (Nuclear Chicago Corp., 17 to 28 c/m-mole), then subjected to a chase for 16 to 20 hr with 10^{-3} M unlabeled uridine.

(d) Isolation of RNA

(i) 28 and 18 s RNA

The two major rRNA species were isolated from purified ribosomal subunits as previously described (Amaldi & Attardi, 1968), unless otherwise specified. In order to eliminate any possible DNA contaminant, the final 18 and 28 s RNA preparations were heated at 100°C for 2 to 3 min in 0·1 × SSC (to denature the DNA), (control experiments showed that this heating step causes only a very small amount of hydrolytic scission of the RNA chains, in agreement with reported observations (Doty, Marmur, Eigner & Schildkraut, 1960)), adjusted to 0·5 M-KCl, and filtered twice through nitrocellulose membranes. In some experiments, the RNA samples extracted from the purified subunits, prior to centrifugation in sucrose gradient in acetate–NaCl buffer, were treated with pancreatic DNase (20 µg/ml.) in Tris–K–Mg for 30 min at room temperature, re-extracted with dodecyl SO$_4$–phenol, precipitated with ethanol, and dissolved in acetate–NaCl buffer.

(ii) Ribosomal RNA precursors

The rRNA precursors were extracted from isolated nucleoli according to the procedure previously reported (Jeanteur et al., 1968), with the following modifications: after the sucrose gradient centrifugation in dodecyl SO$_4$ buffer, the fractions corresponding to the 45 and 32 s peaks were separately pooled, precipitated with 2 vol. of ethanol, redissolved in 0·01 M-acetate buffer (pH 5·0) containing 0·5% dodecyl SO$_4$, and extracted with phenol at 55°C for 3 min (Scherrer & Darnell, 1962) in order to remove any residual protein. After ethanol precipitation, the 45 s material was dissolved in 10^{-3} M-Tris buffer (pH 7·0), $2·5 \times 10^{-4}$ M-EDTA, heated at 80°C for 3 min (negligible degradation of 45s RNA occurs under these conditions (Jeanteur et al., 1968)), quickly cooled to 0°C and run on a 5 to 20% sucrose gradient in the same buffer: the purpose of this step was to separate any possible heterogeneous RNA contaminant (Jeanteur et al., 1968). The 32 s material was either treated in the same way as the 45 s RNA or run on a 15 to 30% sucrose gradient in dodecyl SO$_4$ buffer, sometimes with a subsequent cycle of heat denaturation–sedimentation in low ionic strength. Omission of the heat–low ionic strength treatment did not cause any detectable difference in hybridization behavior of 32s RNA.

(e) Isolation and denaturation of DNA

DNA was extracted from total HeLa cells or from Escherichia coli by the Marmur procedure (Marmur, 1961), with three additional phenol deproteinization steps. The DNA was denatured by addition of NaOH to 0·17 N; after 10 min the pH was brought back to neutrality with HCl. The final measured hyperchromic effect was about 20%. Analysis of the sedimentation behavior of the denatured DNA in a 5 to 20% sucrose gradient in 0·1 N-NaOH revealed a broad band, with a modal sedimentation coefficient of approximately 15 s and more than 90% of the material sedimenting between 9 and 27 s. By applying Studier's formula (Studier, 1965), the molecular weight of the denatured DNA chains was estimated to range between about 0·4 and $9·0 \times 10^6$.

(f) RNA–DNA hybridization and analysis of size and base composition of hybridized RNA

The hybridization experiments were carried out by incubating denatured DNA, at a concentration of 22 µg/ml., and varying amounts of RNA in 2 × SSC at 75°C for 8 hr, unless otherwise specified. The reaction was stopped by quick cooling. Isolation and

132

analysis of the RNA–DNA hybrids were carried out according to the procedure previously reported (Attardi *et al.*, 1965*a*), which was somewhat modified. After pancreatic RNase digestion (5 μg/ml., 1 hr at 22°C), each sample was run through a 0·9 cm × 55 cm Sephadex G100 column equilibrated with 2 × SSC at room temperature. 2-ml. fractions were collected and the O.D.$_{260}$ measured. The fractions containing DNA (the recovery was in general from 80 to 90%) were pooled, adjusted to 0·5 M-KCl–0·01 M-Tris buffer, pH 7·4, and filtered through nitrocellulose membranes (Bac-T-Flex, type B6, Schleicher & Schuell (Nygaard & Hall, 1963)); these were washed with about 150 ml. 0·5 M-KCl–0·01 M-Tris buffer at 60°C. DNA retention by the membrane was frequently tested by measuring the O.D.$_{260}$ of the filtrate, and always found to be at least 95%. The nitrocellulose membranes were either dried and analyzed for radioactivity, or used for determination of size and base composition of the hybridized RNA. For this purpose, the membrane was placed in an inverted position in a 20-ml. beaker containing 1·5 ml. dissociation medium, heated in a boiling water bath (the temperature of the sample was 85 to 90°C) for 2·5 min: the membrane was removed with forceps and rinsed on each face with 0·15 ml. of the dissociation medium. The eluted material (containing more than 85% of the original cts/min) was heated 4 min at 100°C, quickly cooled and run through a 16-ml. 5 to 20% sucrose gradient in PO$_4$–NaCl–formaldehyde buffer in the Spinco SW25.3 rotor at 25,000 rev./min for 36 hr at 4°C. As a control for the size distribution of the original RNA, subjected to the same thermal treatment as the hybridized and dissociated RNA, a portion of the labeled RNA species was incubated in 2 × SSC under the conditions used for the RNA–DNA mixtures, quickly cooled, and diluted 20 times with dissociation medium: 1·5 ml. of the diluted sample was submitted to the same heating steps used in the dissociation of RNA–DNA hybrids and run in a sucrose gradient in PO$_4$–NaCl–formaldehyde buffer as specified above. For base composition analysis of different components of the hybridized and dissociated RNA, fractions corresponding to selected portions of the sucrose gradient pattern were separately pooled, precipitated with 2 vol. of ethanol in the presence of carrier RNA, and subjected to alkali digestion and Dowex 1-X8 chromatography under the conditions previously described (Attardi, Parnas, Hwang & Attardi, 1966). In each hybridization experiment, one or several controls for non-specific background were done by incubating *E. coli* DNA with the labeled RNA under conditions identical to those used in the hybridization with HeLa DNA. This background was always found to be less than 3% of the value obtained with HeLa DNA. All the hybridization values presented in this paper have been corrected for this background.

(g) *Isotope counting procedures*

The nitrocellulose membranes carrying the RNA–DNA hybrids were either glued to planchets and counted in a low-background gas-flow counter (Nuclear Chicago Corporation) or analyzed in a Packard Tricarb liquid-scintillation counter in toluene–POP–POPOP mixture. The fractions of the sucrose gradient runs of dissociated RNA (∼ 0·5 ml. each) were diluted with 1·5 ml. of water, mixed with 20 ml. of Bray's solution (Bray, 1960) and counted in the scintillation counter. The fractions of Dowex 1-X8 chromatography for ^{32}P-labeled nucleotide analysis were dried in glass vials and, after addition of toluene–POP–POPOP, counted in the scintillation counter.

(h) *Quantitative analysis of RNA–DNA hybrids*

The specific activity of the RNA samples was determined by counting in the scintillation counter portions directly plated on nitrocellulose membranes, which had been previously washed with 0·5 M-KCl–0·01 M-Tris buffer, and dried under the same conditions used for membranes carrying the RNA–DNA hybrids. The amount of RNA hybridized in each incubation mixture was determined from the measured radioactivity in the hybrid, on the basis of the specific activity of the RNA and of the amount of DNA recovered after Sephadex chromatography.

For reasons which will be presented in Results section (e), all competition experiments were performed by using a level of labeled RNA and a range of concentrations of competing unlabeled RNA which remained within the rising portion of the DNA saturation curve up to the initial part of the saturation plateau for the RNA species involved. Each experiment included a control set of tubes in which the competition with the labeled RNA by the homologous unlabeled species was tested. The competition to be expected in these homologous mixtures was calculated on the basis of the dilution of labeled by unlabeled molecules and of the change in position on the saturation curve. In the heterologous competition experiments, the fitting of the competition data to a given model was tested as is described below for two particular situations which are relevant to the present study. Given a labeled RNA species S, the theoretical curve of competition for sites in DNA by a larger unlabeled RNA species L containing all the sequences of S was drawn according to the formula:

$$l = \frac{s}{\text{fraction of molecule L represented by S sequences}}$$

where l and s represent the amount of unlabeled L and S, respectively, which give the same level of competition (i.e. the theoretical curve of competition by the L species is displaced on the axis of abscissae with respect to the homologous competition curve by a factor l/s) (Fig. 1). On the other hand, given a labeled species L, the maximum competition to be expected by a smaller unlabeled RNA species S, the sequences of which are contained in L, was calculated by assuming an infinite dilution of the fraction of radioactivity of hybridized L which is due to the S-portion of L, It should be noticed that the calculation of theoretical competition in the two situations considered above was based on the assumption that the hybridization with S sites in DNA of the S-portion of L molecules follows the kinetic behavior of the hybridization of S molecules (see Discussion section (f))

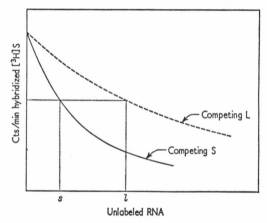

Fig. 1. Theoretical competition for sites in DNA between a labeled RNA species (S) and a larger unlabeled RNA species (L) containing all the sequences of S. See explanation in the text (Materials and Methods section (h)).

3. Results

(a) *Effect of temperature and time of incubation on RNA–DNA hybrid formation*

Figure 2(a) shows the effect of the temperature of incubation on the efficiency of hybridization of HeLa cell 45 s RNA with the homologous DNA, as tested at a low RNA to DNA ratio (~1:110) and for eight hours incubation. It appears that the amount of RNA–DNA hybrids formed increases with the temperature up to 75°C, and thereafter remains constant up to about 85°C. This high optimum hybridization temperature is presumably due to the high melting temperature of rRNA precursors, to be expected from their high (G + C) content (Jeanteur *et al.*, 1968) and their

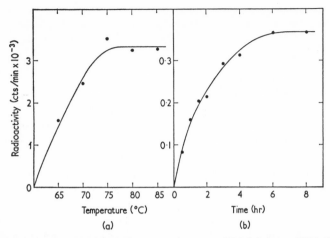

Fig. 2. Effect of temperature and time of incubation on ^{32}P-labeled 45 s RNA–HeLa DNA hybrid formation.

(a) Each incubation mixture contained 25 μg HeLa DNA and 0·23 μg ^{32}P-labeled 45 s RNA in 1·0 ml. After 8 hr incubation at different temperatures, the samples were quickly cooled, and the RNA–DNA hybrids isolated by RNase digestion–Sephadex chromatography, as described in Materials and Methods section (f).

(b) A large incubation mixture containing 22 μg HeLa DNA/ml. and 0·033 μg ^{32}P-labeled 45 s RNA/ml. was incubated at 75°C. At various intervals, 1·0-ml. samples were removed, quickly cooled, digested with 5 μg pancreatic RNase at 22°C for 60 min, adjusted to 0·5 M-KCl–0·01 M-Tris buffer, pH 7·4, filtered through nitrocellulose membranes and washed, as described in Materials and Methods section (f). The zero-time sample was used as a background ($\sim$ 10%).

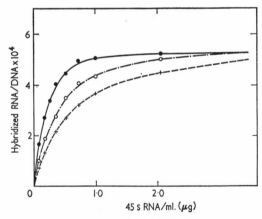

Fig. 3. Effect of time of incubation at 75°C of the hybridization mixture on the shape of the curve of saturation of HeLa DNA by ^{32}P-labeled 45 s RNA. Each incubation mixture contained, in a volume of 1·0 ml., 22 μg DNA and the amount of ^{32}P-labeled 45 s RNA indicated on the axis of abscissae.

—●—●—, 8 hr incubation; —·—O—·—O—·—, 4 hr incubation; —−+—−+—−, 2·5 hr incubation.

high degree of hydrogen bonding (Muramatsu et al., 1966). At 75°C, the level of hybridization reaches a maximum after about six hours incubation (Fig. 2(b)). As expected for a bimolecular reaction (Nygaard & Hall, 1964), with increasing RNA concentration, the time required for maximum hybrid formation decreases, so that at a high RNA input level maximum hybridization occurs already in 2·5 hours (Fig. 3): this behavior results in a marked change in the shape of the saturation curve with the length of incubation, while the final saturation level appears to be constant. For 28 s RNA, maximum hybrid formation was obtained between 70 and 75°C, and, at 75°C, in about four hours incubation. In all the experiments described below, a temperature of incubation of 75°C and a time of incubation of eight hours were used.

(b) Analysis of size and base composition of hybridized RNA after RNase digestion

Figure 4 shows the sedimentation profile of the RNA chains recovered from hybrids between 45 s RNA and HeLa DNA after RNase digestion of non-base-paired segments. At all ratios of RNA to DNA used in the incubation mixtures, there is a fraction of the hybridized and dissociated RNA which overlaps in the sedimentation pattern with the original RNA subjected to the same thermal treatment as the hybrid, and slower sedimenting material, which under the present conditions of sedimentation moves about one-half as fast as the input RNA. The amount of the heavier components reaches a maximum for ratios of input RNA to DNA of about 1:10 and is definitely decreased at a ratio of 1:4; the lighter components augment in proportion to the heavier chains up to an RNA to DNA ratio of about 1:10, and appear to be further increased at a ratio of 1:4. Results similar to those described above, as concerns the size distribution of the RNase-resistant hybridized RNA,

TABLE 1

Nucleotide composition of HeLa 45 s RNA hybridized to homologous DNA, determined after fractionation according to size

| RNA component | Moles % | | | | Purines | $\dfrac{Cp}{Ap}$ |
	Cp	Ap	Up+ψp	Gp	Pyrimidines	
45 s original†	35·6	12·4	16·9	35·1	0·90	2·87
thermally treated	38·1	12·3	18·2	31·4	0·78	3·10
Heavy hybridized 45 s segments	36·7	13·3	19·0	31·0	0·77	2·75
Light hybridized 45 s segments	25·0	22·0	7·9	45·1	2·04	1·14

An incubation mixture, containing 33 μg HeLa DNA and 1·5 μg ^{32}P-labeled 45 s RNA in a total volume of 1·5 ml., was incubated at 75°C for 8 hr, and RNA–DNA hybrids were isolated as described in Materials and Methods section (f). The hybridized RNA was dissociated from DNA and analyzed in a sucrose gradient. The fractions corresponding approximately to the upper third and, respectively, the lower two-thirds of the gradient (see, for reference, the sedimentation patterns of Fig. 4) were separately pooled, precipitated with 2 vol. of ethanol in the presence of carrier RNA, and analyzed for nucleotide composition as previously described (Attardi et al., 1966). As a control, a portion of the ^{32}P-labeled 45 s RNA, subjected to the same thermal treatment as the hybridized RNA, was also analyzed for base composition.

† Jeanteur et al., 1968.

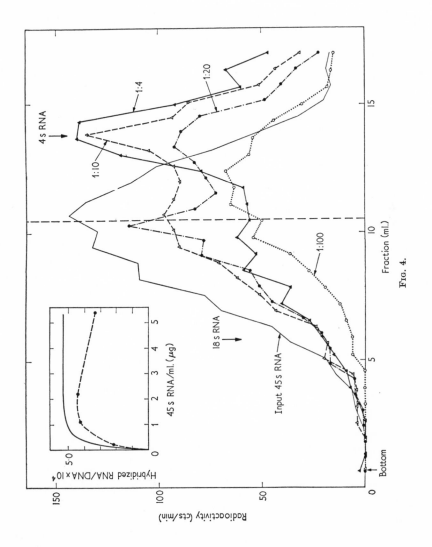

FIG. 4.

relative to that of the original thermally treated RNA, at various RNA to DNA ratios, were obtained for 28 s RNA, in confirmation of earlier observations (Attardi *et al.*, 1965*b*). The heavier chains of hybridized 45 s RNA, on the basis of their relative sedimentation position with respect to 4 s and 18 s RNA markers denatured with formaldehyde (Jeanteur *et al.*, 1968) and run in a sucrose gradient in PO_4–NaCl–formaldehyde buffer under the same conditions as the hybridized and dissociated RNA (Fig. 4), appear to have an average sedimentation coefficient in formaldehyde of about 6 s, corresponding to a length of 450 to 500 nucleotides (Boedtker, 1968). Base composition analysis reveals a striking difference between the light RNA chains and the heavier components recovered from hybrids formed at an RNA to DNA ratio of about 1:20 (Table 1): while the heavier components have a nucleotide composition substantially identical to that of the original 45 s RNA, the lighter chains show a marked increase in purine content.

(c) *Specificity of site recognition by 28 and 18 s RNA*

As a preliminary to the analysis of the relationship between the two major rRNA components and the rRNA precursors, the capacity of 28 s (and 18 s RNA) to discriminate between 28 and 18 s portions of the 45 s DNA sites in the hybridization assay was investigated. No competition for sites in DNA was observed between highly purified 28 and 18 s RNA extracted from isolated ribosomal subunits (Fig. 5); on the contrary, a partial competition was detected between the two rRNA components extracted from the total ribosome–polysome fraction: in this case (as in the previously reported observations (Attardi *et al.*, 1965*b*)), a small degradation of 28 s RNA to molecules sedimenting as 18 s was very probably responsible for the apparent partial competition between the two rRNA species. In all the experiments described in the present work, rRNA components derived from isolated ribosomal subunits were used.

(d) *Saturation experiments*

Figure 6 shows typical DNA saturation curves obtained with 18, 28, 32 and 45 s RNA. It appears that the saturation levels for 18, 28 and 45 s RNA are proportional to the molecular weights of these RNA components (Jeanteur *et al.*, 1968); the ratio of RNA to DNA required to reach saturation of DNA sites is also roughly proportional to the size of the RNA species involved. The behavior of 32 s RNA is at variance with that just described, in that the curve of DNA saturation by this species

---

Fig. 4. Composite diagram showing the sedimentation profile of ^{32}P-labeled 45 s RNA hybridized with HeLa DNA at different RNA to DNA ratios, as compared to that of the original RNA.

Each incubation mixture contained 22 μg HeLa DNA/ml. and varying amounts of RNA in a total volume of 2·5 ml. The hybridized RNA was dissociated from DNA, run in a sucrose gradient and analyzed as described in Materials and Methods sections (f) and (g). A portion of the original ^{32}P-labeled 45 s RNA was subjected to the same thermal treatment as the hybridized RNA and run in a sucrose gradient. The sedimentation positions of 4 s and 18 s RNA markers denatured with formaldehyde (Jeanteur *et al.*, 1968) and run in sucrose gradient in PO_4–NaCl–formaldehyde buffer under the same conditions as the hybridized RNA are indicated by arrows. In each sedimentation profile of hybridized and dissociated RNA, the amount of chains with sedimentation properties similar to those of the input RNA was determined by measuring all radioactivity distributed on the heavy side of the median of the input RNA distribution and doubling it (Attardi *et al.*, 1965*b*).

In the insert, the proportion of total hybridized RNA represented by these heavier chains, for each RNA to DNA ratio, was used to estimate in a typical saturation curve (——) the component pertaining to "complete" hybrids (————).

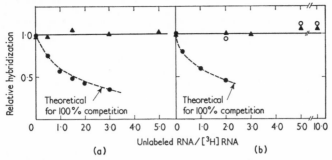

FIG. 5. Tests of competition for sites in HeLa DNA between ^{3}H-labeled 18 s RNA and unlabeled 28 and 18 s HeLa RNA (a), and between ^{3}H-labeled 28 s RNA and unlabeled 28 and 18 s HeLa RNA and 23 s *E. coli* RNA (b).

(a) Each incubation mixture contained 22 μg HeLa DNA/ml., 0·033 μg ^{3}H-labeled 18 s RNA/ml. and variable amounts of unlabeled RNA, in a total volume of 1·5 ml. (▲) 28 s RNA; (●) 18 s RNA.

(b) Each incubation mixture contained 22 μg HeLa DNA/ml., 0·167 μg ^{3}H-labeled 28 s RNA/ml. and variable amounts of unlabeled RNA, in a total volume of 1·5 ml. (▲) 18 s RNA; (●) 28 s RNA; (○) 23 s *E. coli* RNA.

The ratios on the axis of abscissae are expressed on a weight basis. The dashed lines represent the theoretical 100% competition: this was calculated from the saturation curves obtained with the same preparation of ^{3}H-labeled 18 s and ^{3}H-labeled 28 s RNA, respectively, on the basis of the dilution of labeled by unlabeled molecules and of the change in position on the saturation curve.

rises more slowly, with increasing input levels, than the saturation curves of the other species, and only at a relatively high ratio of RNA to DNA ($\sim$1:10) does it appear to be levelling off; furthermore, the presumptive saturation level for 32 s RNA is higher than expected from its molecular weight, though not so high as the level found for 45 s RNA. That this behavior is not due to the presence of contaminating heterogeneous nuclear RNA is suggested by the observation that identical results were obtained with a 32 s preparation more extensively purified by a cycle of heat treatment–sedimentation in low ionic strength buffer (see Materials and Methods section (d) (ii)).

(e) *Tests of competition for sites in DNA between mature rRNA species and rRNA precursors*

Preliminary experiments, aimed at defining the conditions of maximum specificity for the competition experiments between the two major rRNA species and their precursors, indicated that a non-specific interference of 32 s RNA *versus* 18 s RNA occurred when saturating or near-to-saturating amounts of the rRNA precursor were used; likewise, at saturating levels of 45 s RNA, the observed combined competition *versus* 28 and 18 s RNA was greater than could be accounted for by the total ribosomal portion of this precursor (Jeanteur *et al.*, 1968). On the basis of these observations, all the competition experiments to be described below were carried out by using a level of labeled RNA and a range of concentrations of competing unlabeled RNA which corresponded to the rising portion of the respective saturation curves up to the initial part of the saturation plateau. It appears from Figure 7(a) that, under the conditions of hybridization defined above, 32 s RNA does not compete for sites in DNA with ^{3}H-labeled 18 s RNA; on the contrary, it competes with

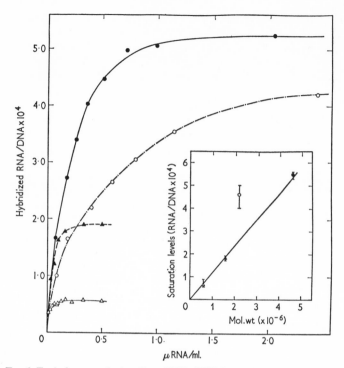

Fig. 6. Typical curves of saturation of HeLa DNA by 18, 28, 32 and 45 s RNA.
[3]H-labeled 18 s RNA and ^{32}P-labeled 28, 32 and 45 s RNA were used in these experiments. Each incubation mixture contained 22 μg HeLa DNA/ml. and the amount of RNA indicated on the axis of abscissae. —△—△—, 18 s; --▲—▲--, 28 s; --○--○--, 32 s; —●—●—, 45 s.

In the insert, the relationship between saturation values (range and average) obtained in different experiments with each RNA species, and its molecular weight is shown. The molecular weights used here were the same as those previously used in the partial sequence studies (see footnote on p. 306).

[3]H-labeled 28 s RNA to the extent to be expected if about 70% of the 32 s molecule is represented by 28 s sequences.† Under the same conditions, the competition of 45 s RNA *versus* ^{3}H-labeled 18 s RNA (Fig. 7(a)) and, respectively, ^{3}H-labeled 28 s RNA (Fig. 7(b)) is consistent with the model according to which an "average" 45 s molecule contains the sequences of *one* 18 s molecule (corresponding to about 13% of its length) and *one* 28 s molecule (corresponding to about 35% of its length).

Experiments carried out in the reverse sense, namely, by testing the hybridization capacity of ^{32}P-labeled 32s or 45 s RNA in the presence of increasing amounts of unlabeled 28 or 18 s RNA, gave results which were qualitatively in agreement with those just discussed (Figs 8 and 9); however, in this case, the competition with ^{32}P-labeled 45s RNA by the two mature rRNA species appeared to be somewhat greater than expected from the model indicated above (Fig. 9). (The anomalous behavior of 32 s RNA in the DNA saturation experiments (Results section (d)) prevented a calculation of the theoretical competition with ^{32}P-labeled 32 s by unlabeled 28 s RNA.) Another abnormality was observed in the competition

† See footnote on p. 128.

experiments between 32 and 45 s RNA: the competition between these two species, whether it was tested with labeled 32 s RNA and unlabeled 45 s RNA or *vice versa*, was found to be greater than expected from their molecular weight ratio, under the assumption, implicit in the previously discussed results, that there is one 32 s segment per 45 s molecule (Fig. 10). The possible explanations for these discrepancies will be discussed below.

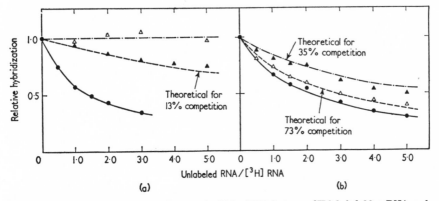

(a) (b)

FIG. 7. (a) Tests of competition for sites in HeLa DNA between ³H-labeled 18 s RNA and unlabeled 18, 32 and 45 s RNA. Each incubation mixture contained 22 μg HeLa DNA/ml., 0·033 μg ³H-labeled 18 s RNA/ml. and variable amounts of unlabeled RNA, in a total volume of 1·5 ml. The dashed line represents the theoretical 13% competition curve, drawn as explained in Materials and Methods section (h) (Fig. 1). (●) 18 s RNA; (△) 32 s RNA; (▲) 45 s RNA.

(b) Tests of competition for sites in HeLa DNA between ³H-labeled 28 s RNA and unlabeled 28, 32 and 45 s RNA. Each incubation mixture contained 22 μg HeLa DNA/ml., 0·033 μg ³H-labeled 28 s RNA/ml. and variable amounts of unlabeled RNA, in a total volume of 1·5 ml. The ratios on the axis of abscissae are expressed on a weight basis. The dashed and dashed–dotted lines represent, respectively, the theoretical 73 and 35 % competition curves, drawn as explained in Materials and Methods section (h) (Fig. 1). (●) 28 s RNA; (△) 32 s RNA; (▲) 45 s RNA.

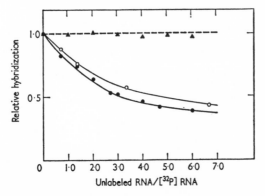

FIG. 8. Tests of competition for sites in HeLa DNA between ³²P-labeled 32 s RNA and unlabeled 32, 28 and 18 s RNA. Each incubation mixture contained, in a volume of 1·0 ml., 22 μg HeLa DNA, 0·05 μg ³²P-labeled 32 s RNA and variable amounts of unlabeled RNA. The ratios on the axis of abscissae are expressed on a weight basis.

(●) 32 s RNA; (○) 28 s RNA; (▲) 18 s RNA.

141

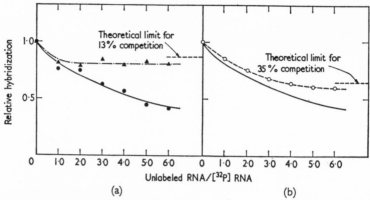

Fig. 9. Tests of competition for sites in HeLa DNA between ^{32}P-labeled 45 s RNA and unlabeled 45 and 18 s RNA (a) or 28 s RNA (b). Each incubation mixture contained, in a volume of 1·0 ml., 22 μg HeLa DNA, 0·05 μg ^{32}P-labeled 45 s RNA and variable amounts of unlabeled RNA. The ratios on the axis of abscissae are expressed on a weight basis. The dashed horizontal lines in (a) and (b) represent the theoretical limits for 13% and 35% competition, respectively, calculated as explained in Materials and Methods section (h).

(a) (●) 45 s RNA; (▲) 18 s RNA. (b) (○) 28 s RNA.

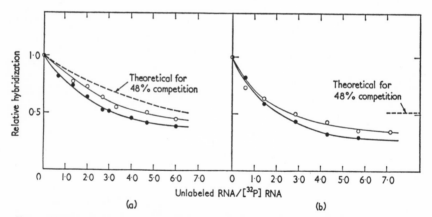

Fig. 10. Tests of competition for sites in HeLa DNA between ^{32}P-labeled 32 s RNA and unlabeled 32 and 45 s RNA (a), and between ^{32}P-labeled 45 s RNA and unlabeled 45 and 32 s RNA (b).

(a) Each incubation mixture contained, in a volume of 1·0 ml., 22 μg HeLa DNA, 0·05 μg ^{32}P-labeled 32 s RNA and variable amounts of unlabeled RNA. (●) 32 s RNA; (○) 45 s RNA.

(b) Each incubation mixture contained, in a volume of 1·0 ml., 22 μg HeLa DNA, 0·175 μg ^{32}P-labeled 45 s RNA and variable amounts of unlabeled RNA. (●) 45 s RNA; (○) 32 s RNA.

The ratios on the axis of abscissae are expressed on a weight basis. The dashed line in (a) represents the theoretical 48% competition and the dashed horizontal line in (b) the theoretical limit for 48% competition, calculated as explained in Materials and Methods section (h).

4. Discussion

The purpose of this work has been to verify by an independent approach the relationship between rRNA and its precursors which has been brought out in the past few years by a considerable amount of kinetic and structural evidence (see review by Perry, 1967; Wagner *et al.*, 1967; Jeanteur *et al.*, 1968; Weinberg *et al.*, 1967; Willems *et al.*, 1968). RNA–DNA hybridization has been used as an analytical tool in the present investigations. The high resolution required for such a type of analysis has made heavy demands both on the purity of the RNA species utilized and on the specificity of the RNA–DNA hybrids investigated.

(a) *Purity of RNA species*

The conditions used for obtaining highly purified 28 and 18 s RNA from ribosomal subunits and rRNA precursors from isolated nucleoli have been previously discussed (Amaldi & Attardi, 1968; Jeanteur *et al.*, 1968). The saturation of DNA sites by relatively low inputs of 28 and 18 s RNA and the base composition data of the hybridized rRNA (Attardi *et al.*, 1965*b*) indicate that the level of the possible contamination of these RNA components by residual labeled mRNA (surviving the chase) was not disturbing in the present experiments. 45 and 32 s rRNA precursors prepared from nucleoli under the conditions used here have been shown to be essentially free of contaminating heterogeneous nuclear RNA. The introduction, in the purification procedure, of a step of heat denaturation and sedimentation in low ionic strength buffer was expected to separate any possible residual RNA of this type (Jeanteur *et al.*, 1968). The finding of a saturation of DNA sites by 45 s RNA at an input RNA to DNA ratio much lower than that required for saturation of DNA by heterogeneous nuclear RNA (Wilkie, Houssais & Attardi, manuscript in preparation) and the base composition data of the hybridized 45 s RNA indicate clearly that heterogeneous nuclear RNA was not involved to any appreciable extent in the RNA–DNA hybridizations analyzed here.

(b) *Recovery of RNA–DNA hybrids*

The large size of the 45 s rRNA precursors and the existing hints of a tandem arrangement of 45 s stretches in DNA (Penman, Vesco & Penman, 1968) raised the possibility that rDNA chains completely base-paired with 45 s RNA might not be retained on the nitrocellulose membranes, as is known to occur for double-stranded DNA segments (Nygaard & Hall, 1963): this possibility seemed to be the more real as the distribution of sedimentation coefficients of the alkali-denatured DNA chains used in the present work was found to correspond to a range of molecular weights from about $4{\cdot}0\times10^5$ to 9×10^6. A selective loss through the nitrocellulose membranes of complete RNA–DNA hybrids would obviously lead to an underestimate of the fraction of HeLa DNA complementary to 45 s precursors. However, the finding that 28 and 18 s RNA, which are expected to hybridize only with portions of the rDNA chains, give DNA saturation levels which are similar, on a molar basis, to the saturation level obtained with 45 s RNA, argues against an appreciable loss of completely base-paired RNA–DNA hybrids involving the latter RNA species. This suggests the presence of non-base-paired regions in the DNA chains involved in hybrid formation.

(c) *Specificity of RNA–DNA hybrids*

The stringent criteria introduced for the analysis of specificity of RNA–DNA hybrids involving rRNA in animal cells (Attardi *et al.*, 1965*b*) have been extended in

143

the present work to the hybrids formed with rRNA precursors. As previously shown for 28 and 18 s RNA, an analysis of the size distribution and base composition of the RNA recovered from hybrids between HeLa DNA and 45 s RNA, after RNase digestion of non-base-paired segments, has revealed the presence of two partially overlapping classes of RNase-resistant chains. One class, represented by the heavier chains, which for input RNA levels up to the beginning of the DNA saturation plateau represents about 80% of the total, has sedimentation properties and base composition similar to those of the original RNA subjected to the same thermal treatment: these heavier components are presumably derived from complexes involving rDNA sites. The regular and complete, or nearly so, base pairing which apparently exists in these complexes, as judged from their RNase resistance, suggests that the hundreds of rDNA sites per cell which are involved in their formation (see below) are similar to each other in sequence. The lighter RNase-resistant chains differ considerably both in size distribution and base composition from the original thermally treated RNA: the marked increase in purine content in these chains suggests that they derive from RNase attack of 45 s molecules imperfectly base-paired with DNA. The nature of the DNA segments involved in these imperfect hybrids is not known; however, the occurrence of these complexes in the form of an apparently distinct class, both as concerns size and base composition (see also Attardi et al., 1965b), suggests that the DNA stretches involved, at least at non-saturating RNA input levels (see below), may not belong to rDNA sites.

The amount of heavier and lighter hybridized RNA chains increases with the input RNA level approximately in parallel, up to a ratio of RNA to DNA corresponding to the beginning of the saturation plateau (about 1:10); at higher ratios of RNA to DNA the amount of heavier components definitely decreases, whereas the lighter chains continue to increase. It seems possible that multiple occupancy of rDNA sites or interactions between 45 s molecules which make them incompletely available for base pairing with DNA are responsible for the decrease in heavier hybridized RNA segments and the increase in lighter components at high RNA to DNA ratios. If this is so, a part of the lighter RNA chains recovered from hybrids formed at high RNA input levels may derive from specific, though incomplete, complexes with rDNA sites. The situation appears to be similar to that observed with 28 s RNA–HeLa DNA hybrids (Attardi et al., 1965b).

(d) Lack of competition for sites in DNA between 28 and 18 s RNA

A prerequisite for the analysis carried out in this work was obviously the capacity of the hybridization assay to discriminate between sequences of 28 and 18s RNA. Previous investigations from this laboratory had indeed indicated the occurrence of a partial hybridization competition between the two rRNA species isolated from the total ribosome–polysome fraction (Attardi et al., 1965b). It seemed likely that the partial competition in these experiments was due to a small degradation of 28 s RNA to molecules sedimenting as 18 s: this degradation was in fact suggested by the somewhat lower 28 to 18 s RNA ratio than expected from the molecular weight ratio (Amaldi & Attardi, 1968). The results of the experiments reported in the present work have supported this interpretation. In fact, no competition for sites in DNA was observed between 28 and 18 s RNA derived from isolated ribosomal subunits, whereas a partial competition was again detected between the two rRNA components extracted from the total ribosome–polysome fraction. The use of isolated ribosomal subunits as a source of highly purified rRNA components proved thus to be essential in this work. The lack of competition for sites in DNA between 28 and 18 s RNA from higher organisms has been previously reported (McConkey & Hopkins, 1964; Ritossa, Atwood, Lindsley & Spiegelman, 1966; Brown & Weber, 1968) and is in agreement with the extensive sequence differences between the two rRNA species (Amaldi & Attardi, 1968).

(e) *DNA saturation levels with rRNA and its precursors*

The levels of saturation of DNA by 18, 28 and 45 s RNA, as measured from the total hybridized RNA retained on the membrane filters, were found in the present work to be proportional to the molecular weights of these RNA species. The fraction of total hybridized RNA which can be recovered as RNA chains of the same size distribution and base composition as the original thermally treated RNA, and which is presumably involved in hybridization with specific sites, is approximately the same (70 to 80% for input RNA levels up to the beginning of the saturation plateau) for 28 s RNA (Attardi *et al.*, 1965b) and 45 s RNA. Thus, it seems reasonable to conclude that the raw saturation levels obtained with these RNA species, and presumably also with 18 s RNA, reflect the fraction of DNA which is represented by the corresponding sites. Therefore, the proportionality between molecular weight of the two rRNA species and their 45 s precursor, on the one hand, and the extent of their homology to DNA, on the other, supports the model according to which an "average" 45 s molecule contains the sequences of one 28 and one 18 s molecule. 32 s RNA behaves abnormally, in that with increasing input levels it approaches saturation of DNA much more slowly than the other RNA species and, furthermore, the presumptive saturation plateau for this RNA is higher than expected from its molecular weight, though lower than the level found for 45 s RNA (see Discussion section (g)).

If the figure of $1 \cdot 2 \times 10^{13}$ daltons is used for the average DNA content of a HeLa cell (McConkey & Hopkins, 1964) and the figure of $4 \cdot 6 \times 10^6$ daltons for the molecular weight of 45 s RNA, it can be calculated that the DNA fraction complementary to 45 s RNA found in the present work is equivalent to about 1400 sites per cell: only about 80% of these, or 1100 sites, however, would correspond to the fraction of the hybrids which can be conservatively considered to be specific on the basis of sedimentation behavior and base composition of the dissociated RNA.

As mentioned earlier, the apparently regular and complete hydrogen bonding of the RNA molecules to rDNA sites in these specific hybrids suggests a considerable similarity in sequence of the sites involved. Therefore, the rRNA cistrons in the human genome probably represent a family of genes with a fairly high degree of precision of sequence repetition (Britten & Kohne, 1968). The different conclusions reached on the basis of measurements of thermal stability by Moore & McCarthy (1968) for the rabbit DNA–rRNA hybrids are probably due to the fact that these authors used for their measurements the whole RNase-resistant hybrids, including, therefore, the class of complexes involving short chains, which are particularly abundant at the high RNA to DNA ratios used in those studies: on the basis of the results presented in the present work and an earlier paper (Attardi *et al.*, 1965b), the involvement of rDNA sites in the formation of these complexes at low RNA input levels is doubtful, and at high input levels may be a trivial consequence of over-crowding.

The raw level of DNA saturation found in the present work for 28 s RNA extracted from purified 50 s ribosomal subunits (corresponding to $\sim 1\cdot8 \times 10^{-4}\,\mu g$ 28 s hybridized per μg DNA) is similar to that previously reported for 28 s RNA extracted from the total ribosome–polysome fraction of HeLa cells (Attardi *et al.*, 1965*b*; Huberman & Attardi, 1967). The raw level of DNA saturation for 18 s RNA extracted from 30 s subunits (corresponding to $\sim 7 \times 10^{-5}\,\mu g$ 18 s hybridized per μg DNA) is, on the contrary, appreciably lower than the value found for 18 s RNA derived from the total ribosome–polysome fraction of HeLa cells; this is not surprising, in view of the above discussed contamination of the 18 s component by splitting products of 28 s RNA occurring when rRNA is extracted directly from ribosomes.

(f) *Hybridization–competition experiments between rRNA and its precursors*

The spurious competition phenomena which were observed between rRNA and its precursors when saturating amounts of RNA were used may have the same mechanism (non-specific interaction with DNA sites, interactions between RNA molecules) which leads in the saturation experiments to a decrease, at high RNA input levels, in the specific complexes involving long RNA stretches. In view of these non-specific interference effects, the competition experiments between rRNA and its precursors were performed by using a range of input RNA concentrations corresponding to the rising portion of the RNA saturation curve up to the initial part of the saturation plateau for the RNA species involved. This required that, in drawing the theoretical competition curves, both the displacement along the DNA saturation curve, accompanying the increase in input RNA, and the dilution of labeled by unlabeled RNA molecules be considered (Fig. 1). Under the optimum conditions defined above, unlabeled 45 s RNA was shown to compete for sites in DNA with labeled 28 and 18 s RNA to the extent expected if about 35 and 13%, respectively, of the precursor molecule were involved: these results provide strong support for the model according to which an "average" 45 s molecule contains the sequences of one 28 s and one 18 s molecule. Similarly, unlabeled 32 s RNA competed for sites in DNA with labeled 28 s RNA to the extent expected if about 70% of the 32 s molecule were represented by 28 s sequences. Competition experiments carried out in the reverse sense, by using labeled 32 or 45 s RNA and unlabeled rRNA components, gave results which were qualitatively consistent with those mentioned above; however, in this case, the competition with 45 s RNA by the two rRNA components appeared to be somewhat greater than expected from their relative molecular weights. The explanation for this discrepancy in behavior may lie in the difference in the kinetics of hybridization between 28 and 18 s RNA, on the one hand, and 28 and 18 s "portions" of precursor molecules, on the other. It is reasonable to assume that molecules of an rRNA species hybridize with the corresponding sites in DNA at a faster rate than the equivalent portions of a precursor which is three or seven times larger (as is the case for 45 s RNA *versus* 28 or 18 s RNA, respectively). The shorter time and lower temperature of incubation required to reach maximum hybrid formation in the case of 28 s RNA as compared to 45 s RNA are in agreement with this assumption. It should be pointed out that for the DNA renaturation reaction the apparent nucleation rate constant per site has been found to be inversely proportional to the square root of the length of the DNA strands involved (Wetmur & Davidson, 1968). There is evidence that RNA–DNA hybrids involving long polynucleotide stretches, once formed, are stable at the temperature of incubation (Nygaard & Hall,

1964). Therefore, if the above discussed assumption on the difference in hybridization rate between rRNA species and their precursors is correct, one would expect that in a mixture, for example, of labeled 45 s RNA and an excess of unlabeled 28 s RNA, the unlabeled 28 s molecules would tend to occupy in a stable form the 28 s sites in DNA faster than the "28 s portions" of the 45 s molecules: this would result in a greater apparent competition than expected on the basis of the displacement along the saturation curve of the "28 s portion" of the 45 s sites and of the dilution of labeled by unlabeled 28 s stretches. The higher is the rate at which the unlabeled rRNA species occupies the available sites relative to both the rate of hybridization of the precursor molecules and the rate of their thermal breakdown, the larger will be this "competition amplification" effect. Therefore, one would expect that an appreciable effect of such nature will be observed at relatively high rRNA inputs and low rRNA precursor concentrations, as was the case in the experiments considered here: under these conditions, in fact, a substantial portion of the sites would quickly become unavailable for hybridization with the labeled precursor molecules. The precise quantitation of this phenomenon would require information concerning the time kinetics of hybridization of the rRNA species and the rRNA precursor involved and the kinetics of thermal degradation of the precursor molecules. In the reverse competition experiments, i.e. those utilizing a labeled rRNA species and an unlabeled precursor, no reduction in competition was observed (see above): this was presumably due to the fact that the labeled rRNA species was present in relatively low concentration and therefore hybridized slowly relative to the rate of thermal breakdown of the precursor molecules.

(g) The case of 32 s RNA

The behavior of 32 s RNA in the saturation experiments suggested that this half of the 45 s molecule also might hybridize, though with reduced efficiency, with the DNA segment homologous to the other half of the 45 s precursor. In agreement with this interpretation are the results of competition tests between 32 and 45 s RNA. 32 s RNA, in fact, was found to compete with 45 s RNA for sites in DNA more than expected on the basis of the molecular weights of these two species, though not up to 100%, both in the experiments using labeled 32 s and unlabeled 45 s RNA and in the reverse type of experiments (in the latter case, the "competition amplification" effect discussed above presumably contributed to the results). These observations, on the one hand, agree with the other hybridization results of the present work which argue against the possibility that 45 s RNA is a dimer of 32 s RNA (a possibility which was compatible with the base sequence data (Jeanteur et al., 1968)), and on the other hand, point to a sequence similarity between a portion of the 32 s stretch and a portion of the other half of the 45 s molecule: since the hybridization assay used here could discriminate clearly between 28 and 18 s segments, this sequence similarity must concern the non-ribosomal sequences. The capacity of the non-ribosomal portion of 32 s to hybridize with the DNA segment homologous to the non-ribosomal portion of the other half of the 45 s molecule is indeed not surprising, in view of the remarkable similarity in base composition and partial sequence distribution which has been shown to exist between the two non-ribosomal stretches (Jeanteur et al., 1968). There is another possibility which could be considered in connection with the anomalous behavior of 32 s RNA, namely that this RNA may be slightly contaminated by 41 s RNA, which is an intermediate in the

147

processing of rRNA precursors (Weinberg *et al.*, 1967): this minor nucleolar RNA species would be expected, in fact, to hybridize with a larger stretch of rDNA than 32 s RNA. However, it should be noticed that in the present work the 32 s RNA was purified by two cycles (or even, with identical results, three cycles) of sucrose gradient centrifugation, with only the central portion of the 32 s peak being used at each step. It seems unlikely therefore that any residual contamination by 41 s RNA may account by itself for the large deviations from the expected behavior of 32 s RNA, observed already at relatively low inputs, in the saturation and competition experiments.

(h) *Conclusion*

The analysis carried out in the present work on the relationship between rRNA and its precursors in HeLa cells has provided support by an approach different from those previously used for the model according to which 45 s RNA is the precursor of *both* 28 and 18 s RNA, and 32 s RNA is precursor of 28s RNA *only* (Scherrer *et al.*, 1963; Perry, 1965; Penman, 1966; Greenberg & Penman, 1966; Zimmerman & Holler, 1966,1967; Wagner *et al.*, 1967). The experiments described above have not given information about the distribution of 28 and 18 s polynucleotide stretches among the 45 s molecules. However, the available evidence concerning the arrangement of 28 and 18 s cistrons in *Xenopus* (Brown & Weber, 1968; Birnstiel, Speirs, Purdom, Jones & Loening, 1968) and in *Drosophila* (Quagliarotti & Ritossa, 1968) suggests that the 28 and 18 s sequences are contained in the *same* precursor molecule. The quantitative results obtained in the present saturation and competition experiments have furthermore confirmed the molecular weight data (McConkey & Hopkins, 1969) and the base composition and sequence data (Amaldi & Attardi, 1968; Jeanteur *et al.*, 1968; Weinberg *et al.*, 1967; Vaughan *et al.*, 1967; Willems *et al.*, 1968; Roberts & D'Ari, 1968) indicating that each 45 s RNA molecule contains at most the sequences of one 28 and one 18 s RNA molecule, and that only about 70% of the 32 s RNA molecule can be accounted for by sequences of mature rRNA: this implies that about 50% of the sequences in the 45 s molecule and about 30% of the sequences in the 32 s molecule are of non-ribosomal type and are not conserved in the conversion to mature rRNA.

Both the abnormal hybridization behavior of 32 s RNA and the previously published partial sequence data (Jeanteur *et al.*, 1968) suggest that the non-ribosomal portion of 45 s RNA may contain repetitive sequences. These observations, in addition to the unusually high (G + C) content (77%) and low A content (8%) of this non-ribosomal portion, support the idea that it may have a transient structural (rather than informational) role during the processing of the rRNA precursors (Jeanteur *et al.*, 1968). The precise anatomy of the 45 s molecule, that is, the relative arrangement of the 28 and 18 s stretches and of the non-ribosomal stretch(es) remains to be established.

The availability, in a highly purified form, of a set of well-defined inter-related RNA species has provided in the present work a valuable opportunity for investigating some of the problems involved in the identification and titration of specific DNA sequences in the eukaryotic genome. The high resolution achieved in the present experiments is a further confirmation of the specificity exhibited, under proper conditions, by RNA–DNA hybrids involving rDNA in higher organisms.

These investigations were supported by a research grant from the U.S. Public Health Service (GM-11726). One of us (P. J.) was supported by a fellowship from the American Cancer Society while on leave of absence from the Institut National de la Santé et de la Recherche Médicale, France. We are grateful to Mrs L. Wenzel and Mrs B. Keeley for their invaluable help.

REFERENCES

Amaldi, F. & Attardi, G. (1968). *J. Mol. Biol.* **33**, 737.
Attardi, G., Huang, P. C. & Kabat, S. (1965a). *Proc. Nat. Acad. Sci., Wash.* **53**, 1490.
Attardi, G., Huang, P. C. & Kabat, S. (1965b). *Proc. Nat. Acad. Sci., Wash.* **54**, 185.
Attardi, G., Parnas, H., Hwang, M-I.H. & Attardi, B. (1966). *J. Mol. Biol.* **20**, 145.
Birnstiel, M., Speirs, J., Purdom, I., Jones, K. & Loening, U. E. (1968). *Nature,* **219**, 454.
Boedtker, H. (1968). *J. Mol. Biol.* **35**, 61.
Bray, G. A. (1960). *Analyt. Biochem.* **1**, 279.
Britten, R. J. & Kohne, D. E. (1968). *Science,* **161**, 529.
Brown, D. D. & Weber, C. S. (1968). *J. Mol. Biol.* **34**, 681.
Doty, P., Marmur, J., Eigner, J. & Schildkraut, C. (1960). *Proc. Nat. Acad. Sci., Wash.* **46**, 461.
Gilbert, W. (1963). *J. Mol. Biol.* **6**, 389.
Greenberg, H. & Penman, S. (1966). *J. Mol. Biol.* **21**, 527.
Huberman, J. & Attardi, G. (1967). *J. Mol. Biol.* **29**, 487.
Jeanteur, P., Amaldi, F. & Attardi, G. (1968). *J. Mol. Biol.* **33**, 757.
Marmur, J. (1961). *J. Mol. Biol.* **3**, 208.
McConkey, E. H. & Hopkins, J. W. (1964). *Proc. Nat. Acad. Sci., Wash.* **51**, 1197.
McConkey, E. H. & Hopkins, J. W. (1969). *J. Mol. Biol.* **39**, 545.
Moore, R. L. & McCarthy, B. J. (1968). *Biochem. Genetics,* **2**, 75.
Muramatsu, M., Hodnett, J. L. & Busch, H. (1966). *J. Biol. Chem.* **241**, 1544.
Nygaard, A. P. & Hall, B. D. (1963). *Biochem. Biophys. Res. Comm.* **12**, 98.
Nygaard, A. P. & Hall, B. D. (1964). *J. Mol. Biol.* **9**, 125.
Penman, S. (1966). *J. Mol. Biol.* **17**, 117.
Penman, S., Vesco, C. & Penman, M. (1968). *J. Mol. Biol.* **34**, 49.
Perry, R. P. (1962). *Proc. Nat. Acad. Sci., Wash.* **48**, 2179.
Perry, R. P. (1965). *Nat. Cancer Inst. Monogr.* **18**, 325.
Perry. R. P. (1967). In *Progress in Nucleic Acid Research and Molecular Biology*, vol. 6, p. 219. New York: Academic Press.
Quagliarotti, G. & Ritossa, F. M. (1968). *J. Mol. Biol.* **36**, 57.
Rake, A. V. & Graham, A. F. (1964). *Biophys. J.* **4**, 267.
Ritossa, F. M., Atwood, K. C., Lindsley, D. L. & Spiegelman, S. (1966). *Nat. Cancer Inst. Monogr.* **23**, 449.
Roberts, W. K., D'Ari, L. (1968). *Biochemistry,* **7**, 592.
Scherrer, K. & Darnell, J. E. (1962). *Biochem. Biophys. Res. Comm.* **7**, 486.
Scherrer, K., Latham, H. & Darnell, J. E. (1963). *Proc. Nat. Acad. Sci., Wash.* **49**, 240.
Studier, F. W. (1965). *J. Mol. Biol.* **11**, 373.
Vaughan, M. H., Soeiro, R., Warner, J. R. & Darnell, J. E. (1967). *Proc. Nat. Acad. Sci., Wash.* **58**, 1527.
Wagner, E. K., Penman, S. & Ingram, V. M. (1967). *J. Mol. Biol.* **29**, 371.
Weinberg, R. A., Loening, U., Willems, M. & Penman, S. (1967). *Proc. Nat. Acad. Sci., Wash.* **58**, 1088.
Wetmur, J. G. & Davidson, N. (1968). *J. Mol. Biol.* **31**, 349.
Willems, M., Wagner, E., Laing, R. & Penman, S. (1968). *J. Mol. Biol.* **32**, 211.
Zimmerman, E. & Holler, B. (1966). *Fed. Proc.* **25**, 646.
Zimmerman, E. & Holler, B. (1967). *J. Mol. Biol.* **23**, 149.

Separation of Ribonucleic Acids by Polyacrylamide Gel Electrophoresis

ULRICH GROSSBACH AND I. BERNARD WEINSTEIN

Column electrophoresis on polyacrylamide gel (1, 2) has become an invaluable technique for the separation of proteins. The molecular sieving action of the gel also allows the electrophoretic separation of molecules which differ in size but not in charge. Taking advantage of this feature, Richards and Gratzer (3, 4) developed a method for the separation of low molecular weight RNA on acrylamide columns. The method has been used for the purification of transfer RNA from other low molecular weight RNA and for studying the partial degradation of ribosomal RNA by ribonucleases (5–7).

In this communication, polyacrylamide gel electrophoresis of RNA molecules which span a wider range of molecular weights, including those of ribosomal origin, is described. The separation of high molecular weight RNA was achieved by employing gels of low acrylamide concentrations and therefore large pore sizes. The separation of classes of RNA was further enhanced by using columns which had successive layers of gels that contained decreasing concentrations of acrylamide.

MATERIALS AND METHODS

RNA, prepared by extraction with phenol-sodium dodecyl sulfate (8) of larvae of *Chironomus tentans* or adult *Drosophila*, was kindly supplied by Dr. C. Pelling. These preparations contained less than 5% DNA. The sedimentation coefficient of the contaminating DNA was in the range of 17–24S. *Escherichia coli* B sRNA was obtained from General Biochemicals Corp. Yeast RNA and crystalline pancreatic ribonuclease were purchased from Serva (Heidelberg). Acrylamide, N,N'-methylenebisacrylamide, and N,N,N',N'-tetramethylethylenediamine were obtained from Canal Industrial Corp. Dimethylaminopropionitrile was obtained from Serva and gallocyanin from Chroma (Stuttgart).

Electrophoresis was performed in glass tubes of 0.5 cm i.d. and 6.5 or 10 cm length. The discontinuous buffer system of Williams and Reisfeld (9) was used, in which the separation gel contains a tris-HCl buffer of pH 7.5 and the stacking gel a tris-HCl buffer of pH 5.5. The electrode buffer was diethylbarbituric acid-tris pH 7.0. Acrylamide gels of different pore sizes were polymerized from solutions containing monomer concentrations of 2.5, 2.75, 3, 3.1, 3.2, 3.3, 3.4, 3.5, 4, 5, 7, or 10%. The 5, 7, and 10% gels were made according to the instructions supplied with the electrophoresis apparatus (Canal Industrial Corp.). The less concentrated gels were prepared as described by Hjertén (10) and Hjertén et al. (11) and contained an amount of N,N'-methylenebisacrylamide which was 5% of the total monomer concentration. Throughout the remainder of this paper "% gel" refers to the final concentration of acrylamide in the case of 5, 7, and 10% gel and to the total concentration of both monomers in the case of the less concentrated gels. In place of the usual layer of stacking gel, the separation gel was overlaid with 0.2 ml of stacking gel buffer containing 20% sucrose. The sample (5–100 μg of RNA) in 10–200 μl of 20% sucrose, 0.01 M acetate buffer, pH 5.1, and 10^{-4} M $MgCl_2$, was layered on top of the column after mixing with an equal volume of double strength stacking gel buffer. Electrophoresis was carried out at 4°C for 1 hr at 1 mA per tube and then at 2 mA per tube for another 1–3 hr.

After electrophoresis the gels were extruded from the glass tubes by forcing a stream of water through a long No. 20 needle which was carefully inserted between the gel and the glass tube. Gels of 2.5% monomer concentration had to be handled with considerable care. Fixation was performed overnight in 15% acetic acid containing 1% lanthanum acetate (4). The gels were then stained for 24 hr in freshly prepared gallocyanin-chromalum solution (pH 1.6) as described by de Boer and Sarnaker (12) for histochemical studies. Destaining was achieved by rinsing in lanthanum acetate-acetic acid for two to three days. The gels were placed in a trough consisting of a pane of optical glass and a 1 cm high frame of Plexiglas glued to it. The trough was filled with lanthanum acetate-acetic acid, and the gels were photographed using Kodak Wratten filter No. 25 and diffuse illumination from beneath the gel.

EXPERIMENTAL

Determination of RNA on Acrylamide

The complex of RNA and gallocyanin-chromalum has been reported to follow Beer's law under certain conditions (13). Two types of experiments were performed to determine whether gallocyanin-chromalum could be used to quantitate the amount of RNA in acrylamide gels. In the first type of experiment an acrylamide gel was prepared which contained a linear concentration gradient of *E. coli* RNA. This was done by placing a 7% gel solution, which contained 20% sucrose, into the mixing chamber of a device used for preparing linear sucrose gradients. The second chamber contained *E. coli* sRNA (4 mg/ml) dissolved in 7% gel solution. The gradient was delivered at 4°C into a series of glass columns (5 mm i.d.) and the material then allowed to polymerize at room temperature. Immediately after polymerization the gel was extruded, fixed, stained, and rinsed as described under "Methods." The gel was then placed in a trough of optical glass. The trough was filled with lanthanum acetate-acetic acid, and the gel was scanned in a Joyce-Loebl microdensitometer equipped with an orange filter (Ilford 607). The results (Fig. 1) indicated that above a certain minimum concentration of RNA there was an approximately linear relationship betwen RNA concentration and absorbancy.

Yeast RNA was used in the second type of experiment. Test solutions

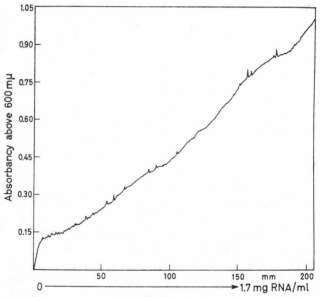

Fig. 1. Absorbancy of the complex of soluble RNA with gallocyanin-chromalum. A column of acrylamide gel was prepared which contained a linear concentration gradient of *E. coli* RNA. The gel was fixed in 15% acetic acid containing 1% lanthanum acetate and stained with gallocyanin-chromalum solution, pH 1.6. The absorbancy above 600 mμ of the RNA-dye complex in the gel was scanned in a microdensitometer equipped with an Ilford filter No. 607. Additional details are given in the text.

152

of 7% separation gel containing from 0.05 to 1.25 mg/ml RNA were prepared. Beginning with the lowest RNA concentration, approximately 0.2 ml quantities of these solutions were successively polymerized over each other in the previously described glass tubes. Both ends of the tube contained a layer of gel without RNA and these regions were used to determine blank values. Immediately after polymerization the gels were extruded, fixed, and stained. It was not possible to scan these gels directly in the densitometer because they consisted of short segments which had a somewhat concave surface. They were therefore photographed in the glass trough using Polaroid transparent film 46-L, a Kodak Wratten filter No. 25, and diffuse illumination from beneath the gel. The image of a gray wedge which had gradations in absorbance of 0.15 was included on each photograph. This made it possible to determine whether absorption was within the linear range of the blackening curve of the photographic film. The photographs of the gels were then scanned in the microdensitometer. To allow correction for uneven illumination, the absorption of the background was recorded as close to each gel as possible. Parallel to this tracing, a line was drawn through parts of the curve recording unstained regions of the gel. The mean values of the pen excursion for

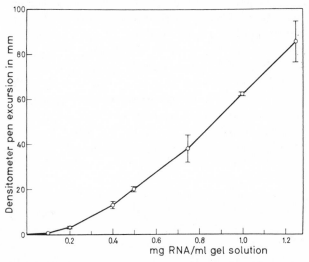

Fig. 2. Absorbancy of the complex of yeast RNA with gallocyanin-chromalum. Short lengths of acrylamide solutions containing increasing concentrations of RNA were polymerized over each other in glass tubes. The gels were fixed and stained as indicated in Figure 1. They were photographed through a Kodak Wratten filter No. 25 using Polaroid film 46-L and a density reference system. The photographs were then scanned in the microdensitometer. The points represent mean values and standard deviations of five replicates. Additional details are given in the text.

five columns were plotted against the RNA concentration (Fig. 2). It can be seen that, in the range of 0.2 to 1.2 mg RNA/ml gel solution, there was an approximately linear relationship between RNA concentration and absorbance.

In subsequent studies both the above methods were used for evaluating relative amounts of RNA on acrylamide gels. The photographic procedure (Polaroid film 46-L or 55 P/N) was used for columns which contained more than one layer of gel, because the beam of the densitometer was deflected at the boundary between gel layers, thus producing an artificial peak.

Fractionation of RNA

1. *Staining with pyronine:* During electrophoresis, the major components of an RNA preparation could be observed as sharp diffraction lines migrating toward the anode. In order to follow the separation in greater detail and to determine the most suitable combination of gel concentrations, pyronine (0.05%) was added to the sample prior to electrophoresis (Kozłowska and Miczyński, personal communication). Pyronine from Merck (No. 7517) was found to be suitable, while samples from several other sources contained positively charged components. Free pyronine could be seen as a brown disc traveling at the salt front and served as a tracking dye. The dye stained RNA red and did not influence the separation pattern. This red color disappeared during fixation and did not interfere with the gallocyanin stain.

2. *Gel concentration:* In 2.5% acrylamide, *Drosophila* and *Chironomus* RNA were separated into three major fractions. To identify these fractions, 4S, 18S, and 28S RNA were obtained from sucrose gradients and each was electrophoresed under identical conditions. The fastest moving band was 4S RNA, the next fastest was 18S, and the slowest moving band was 28S RNA. Similar results were also obtained with 4S, 18S, and 28S RNAs obtained from rat liver.

Electrophoretic resolution was higher than that obtained by sucrose gradient centrifugation, since RNA taken from a peak region of the sucrose gradient gave rise to two fractions on electrophoresis. The RNA from the 28S peak of a sucrose gradient was separated into a major band as well as a minor band, the latter corresponding in position to 18S RNA. Similarly, RNA from the 18S peak of the sucrose gradient revealed on electrophoresis a contaminant corresponding in position to 28S RNA.

Separation was further enhanced by employing columns in which gel solutions of different monomer concentrations (2.5 to 10%) were layered over each other. When RNA obtained from peak regions of a sucrose gradient was then electrophoresed through these layers it was found that:

154

(*1*) 28S RNA migrated into 2.5% gel, barely entered 2.75% gel, and was excluded from 3.3% gel, (*2*) 18S RNA was excluded from 4% gel, and (*3*) 4S RNA traveled with the salt front in less than 7% gels whereas in 7 or 10% gels it migrated somewhat slower than the front.

While traveling through the acrylamide column, an RNA zone became more concentrated when it reached the surface of a gel of lower pore size. In columns composed of a series of gels in which the RNA zone was increasingly retarded, this concentration step was repeated at each boundary, while spreading occurred during migration between the boundaries. The sharpest band was obtained when a zone was allowed to run onto the boundary of a gel from which it was entirely excluded. If the pore size was in a critical range, however, an RNA fraction would concentrate as a fine disc at the gel surface, but, if the voltage gradient was further maintained, part of this RNA migrated into the gel producing a diffuse smear. This phenomenon probably explains why by exploring gel concentrations in the range of 3–4% it was not possible to find a gel concentration which entirely excluded 28S RNA from *Chironomus* or *Drosophila* without giving rise to artifacts in the migration of 18S RNA. For the same reason we did not try to separate RNA on continuous acrylamide gradients.

3. *Fractionation of RNA from whole tissue:* As a result of the observations described above, a combination of 2.5 and 7% gel was selected for the separation of RNA extracted from whole tissue. Stacking gel solution and sample (see "Methods") were layered over each other on top of a column which contained a bottom layer (4–5 cm) of 7% gel and an upper layer (1.5 cm) of 2.5% gel. A typical separation of *Chironomus* RNA is shown in Figure 3 together with its densitometer tracing. In addition to the three major fractions, several minor components were found repeatedly. These were not detected in the UV absorbancy pattern of sucrose gradients of the same RNA samples (kindly run by Dr. C. Pelling). Because of the results obtained with ribosomal and soluble RNA it seems reasonable to assume that the positions of the minor fractions are also related to their sedimentation coefficients

The RNA nature of the minor bands was established by extensive treatment of *Chironomus* RNA with crystalline pancreatic ribonuclease. Gel electrophoresis of these samples revealed only a band which migrated faster than 4S RNA. It appears, therefore, that the bands separated and stained in the present method are RNA fractions since they were degraded to low molecular weight material by extensive treatment with ribonuclease.

The relatively large amount of RNA regularly found in the position of the 4S fraction remains unexplained. This fraction is, however, not homo-

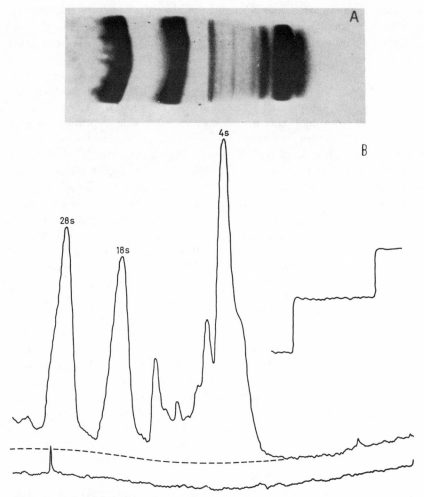

Fɪɢ. 3. Polyacrylamide gel electrophoresis of 50 μg RNA from *Chironomus tentans:* (A) Polyacrylamide gel stained with gallocyanin-chromalum solution, pH 1.6; cathode on the left. (B) Densitometer tracing of photograph of gel shown in A. On the same sheet of graph paper and with identical instrument parameters, the absorption of a reference system was recorded. Part of this tracing is shown on the right. The image of the reference system was included in the photograph. It shows gradations in absorbance of 0.15. Additional details are given in the text.

geneous. It could be separated into several bands when electrophoresed in a 10% gel.

Some evidence was obtained suggesting the existence of additional minor components in *Drosophila* RNA. In experiments with the discontinuous buffer system of Ornstein (1) and Davis (2) (pH 6.9–8.9), four to five very fine but very sharp red bands were regularly observed migrating between the two ribosomal RNA fractions of pyronine-stained samples. These bands disappeared a few minutes after they had been separated from the ribosomal RNA fractions. They could not be seen in gels stained with gallocyanin-chromalum. These bands were not observed in the discontinuous buffer system (pH 5.5–7.5) used in most of our studies. The buffer system of Ornstein (1) and Davis (2) may be useful, therefore, for separating certain minor fractions of RNA.

DISCUSSION

It seems likely that the separation of RNA achieved by the present method of gel electrophoresis is based primarily on differences in size and shape of the RNA fractions, since the types of RNA studied probably have a similar charge to mass ratio. With ribosomal and soluble RNA, the rate of migration through the gel is inversely related to the sedimentation coefficient of the RNA. The molecular sieving action of the acrylamide gel was also apparent when a sample was allowed to travel through succesive layers of gels containing increasing concentrations of acrylamide and hence decreasing pore sizes. RNA fractions migrating as one band in a gel of a certain pore size became separated into several components on entering a gel of smaller pore size and were eventually excluded from very concentrated gels.

In addition to providing sharper resolution of 4S, 18S, and 28S RNA than obtainable by the sucrose gradient technique, the present method has revealed the presence of several minor RNA components. The nature of these minor components remains to be determined. Some of these bands may simply represent products of partial enzymic degradation (5–7). Others, however, may represent specific classes of RNA not previously resolved by conventional procedures. The concentrations of some of these minor components appear to be several orders of magnitude below those of ribosomal RNA and would, therefore, be in the range expected for messenger RNA's. It may be rewarding, therefore, to use gel electrophoresis to study the distribution of radioactivity in RNA obtained from pulse-labeled cells.

Richards and co-workers (4) used acridine orange for staining RNA on acrylamide gels. It is known, however, that the binding of acridine orange by RNA is heterogeneous and depends on the base composition (14). Gallocyanin-chromalum, on the other hand, combines with the phosphate groups of nucleic acids (15) and is, therefore, likely to give uniform staining with different RNA types. The present study demonstrates that gallocyanin-chromalum is a sensible and reproducible stain

157

for detecting and quantitating RNA after electrophoresis in acrylamide gels. Evidence has been presented that the RNA-dye complex follows Beer's law over a reasonable range of RNA concentration. The present procedure must be considered only semiquantitative, however, since the influence of dye concentration, staining, and destaining time, and of stray light during the photographic technique, has not been studied in great detail. An obvious source of error is the different degree of swelling in acetic acid of gels of different concentration. Measurements are strictly comparable only within gel regions of uniform monomer concentration. This factor may explain why in the results presented in Figure 3 the absorption values obtained for ribosomal RNA components are relatively low when compared to the faster components. The differences in gel swelling are, however, reproducible, and correction factors could, therefore, be introduced.

Electrophoresis of RNA on acrylamide gel columns may be a useful analytical technique for the study of minor RNA components. On a larger scale, it might be the method of choice for the preparative purification of RNA species. Preliminary experiments in capillary columns (16) have shown that the method is also suitable for RNA separation on the scale of 10^{-9} gm and might possibly be further reduced. This should greatly facilitate the fractionation of RNA's obtained from single giant cells or cell organelles.

SUMMARY

1. A method is described for the fractionation of RNA from whole tissue by acrylamide gel electrophoresis in a discontinuous buffer system of pH 5.5–7.5. The highest resolution was achieved when the sample migrated through layers of gels of increasing acrylamide concentration and correspondingly decreasing pore size.

2. RNA fractions in acrylamide gel were stained with gallocyanin-chromalum. Relative amounts were determined by densitometry. With *E. coli* sRNA or yeast RNA, the RNA-dye complex in acrylamide gel was shown to follow Beer's law in the range between 0.4 and 1.7 mg RNA/ml.

3. In addition to the two ribosomal and the soluble RNA fractions, several minor components were regularly found in RNA extracted from *Drosophila* and *Chironomus tentans* by the phenol-SDS method. These could not be detected in the UV absorbancy pattern of sucrose gradients of the same preparations.

ACKNOWLEDGMENTS

During the course of these studies we learned from Dr. U. Loening of his method of RNA electrophoresis. We wish to thank him for helpful discussions. We are also indebted to Professor W. Beermann for his kind support and to Dr. C. Pelling for providing RNA samples and for stimulating discussions. We would like to thank Mr. E. Freiberg for carefully drawing the figures and Miss L. Dentzer for her excellent assistance.

REFERENCES

1. ORNSTEIN, L., *Ann. N. Y. Acad. Sci.* **121**, 321 (1964).
2. DAVIS, B. J., *Ann. N. Y. Acad. Sci.* **121**, 404 (1964).
3. RICHARDS, E. G., AND GRATZER, W. B., *Nature* **204**, 878 (1964).
4. RICHARDS, E. G., COLL, J. A., AND GRATZER, W. B., *Anal Biochem.* **12**, 452 (1965).
5. MCPHIE, P., HOUNSELL, J., AND GRATZER, W. B., *Biochemistry* **5**, 988 (1966).
6. GOULD, H., *Biochemistry* **5**, 1103 (1966).
7. GOULD, H., *Biochim. Biophys. Acta* **123**, 441 (1966).
8. HIATT, H. H., *J. Mol. Biol.* **5**, 217 (1962).
9. WILLIAMS, D. E., AND REISFELD, R. A., *Ann. N. Y. Acad. Sci.* **121**, 373 (1964).
10. HJERTÉN, S., *Arch. Biochem. Biophys.*, Suppl. 1, 147 (1962).
11. HJERTÉN, S., JERSTEDT, S., AND TISELIUS, A., *Anal. Biochem.* **11**, 211 (1965).
12. DE BOER, J., SARNAKER, R., *Med. Proc. (South Africa)* **2**, 218 (1956).
13. SANDRITTER, W., DIEFENBACH, H., AND KRANTZ, F., *Experientia* **10**, 210 (1954).
14. BEERS, R. F., AND ARMILEI, G., *Nature* **208**, 466 (1965).
15. PEARSE, A. E. G., "Histochemistry," 2nd ed., p. 210. Little, Brown, Boston, 1961.
16. GROSSBACH, U., *Biochim. Biophys. Acta* **107**, 180 (1965).

UNSTABLE REDUNDANCY OF GENES FOR RIBOSOMAL RNA*

By F. M. Ritossa

The present work was begun with the aim of understanding the mechanism of reversion of the *bobbed* phenotype in *Drosophila melanogaster*. We have previously shown that this phenotype is due to a partial deficiency of DNA complementary to ribosomal RNA (rRNA).[1] We know that the wild-type *bb*+ locus, which probably is cytologically identifiable with the nucleolus organizer, carries at least 130 genes for ribosomal RNA.[2] Much evidence[2, 3] points to the possibility that these genes are clustered and possibly adjacent. If these multiple copies are tandemly arranged, unequal crossing-over can be expected to occur. This was indeed the mechanism we invoked to explain the frequent appearance of *bb* mutations and the reversion of the *bb* phenotype which is paralleled by gain of genes for rRNA.[1, 4] The multiplicity of the rRNA genes was explained on the basis of the comparatively high demand for ribosomal RNA.[2] The possibility that this might not be enough to satisfy all requirements was explored by comparing a variety of tissues which differed widely in their rates of ribosome synthesis. The corresponding DNA preparations showed, however, the same redundancies within experimental error.[4] A striking exception was discovered by Brown[5] in the amphibian oöcyte which exhibited a marked extrachromosomal redundancy. We have here then an example of a new type of gene regulation analogous to the one proposed some years ago[6] that involved the induced production of extrachromosomal copies of specific genes (plasmagenes).

It was of some interest to see whether evidence for a similar mechanism could be detected in *Drosophila*. As an initial attempt a search was made for abrupt changes in rDNA content in stocks carrying the *bobbed*-type deletion. Schalet[7] has shown that crossing-over in the *bb* locus can be of the unequal type and that the frequency of crossing-over within the *bb* region is of the order of 0.6 per cent. One should then expect *bb* mutations to appear in a *bb*+ population with a frequency not higher than this value. Similarly, a population of *bb* individuals should also show *bb*+ revertants with a frequency not higher than 0.6 per cent. The possible operation of an extrachromosomal amplification mechanism would be signaled by a restoration of the rDNA deficiency in a *bobbed* stock at a rate much faster than could be accounted for by the observed crossing-over frequency.

It is the purpose of this paper to present data showing that reversion of the *bb* phenotype can, in certain cases, occur suddenly by the accumulation of genes for ribosomal RNA which, although normally inherited through it, may not be perfectly integrated into the chromosome.

Materials and Methods.—*Drosophila stocks:* g^2ty/y f:= flies were from the Pasadena collection; $In(1)sc^{4L,8R}$ came originally from the Oak Ridge collection; w^asn bb/Y^{-bb} and y v f:= and $In(1)sc^8/y$ f:= Y^{Bs} stocks are from the Bowling Green collection; X $Y^L.Y^s$ (108–9), $y^2su\text{-}w^aw^aY^L.Y^s/y$ v $bb,'O$ is from the University of Rome collection.

Hybridization procedure: P³²- or H³-labeled ribosomal RNA from wild-type larvae of *Drosophila melanogaster* was extracted and purified as previously described.[1, 2] DNA was

160

extracted from adult flies only, according to our standing procedure,[1, 2] and rRNA/DNA hybrids were made on nitrocellulose membrane filters.[8] RNase digestion (20 μg/ml at 30°C for 1 hr in 2 × SSC) was always made after hybridization.

Results.—I found that males of the g^2ty/Y and $y f. = $ /Y stock from Pasadena carry a strong *bb* isoallele in the X, though the mutation is not recorded in the stock list. When this X, which can hereafter be written $g^2ty\ bb$, is combined with an sc^4sc^3 chromosome which carries no rDNA, or with an X carrying a bb^l mutation, the females obtained (females Gl of Fig. 1) exhibit an extreme *bb* phenotype. One of the relevant characteristics of these females is that they are almost completely sterile and the few eggs they lay are dechorionated; they produce no progeny.

If the $g^2\ ty\ bb$ chromosome is combined with a Y^{-bb} chromosome, which is functionally equivalent to a *bb* deficiency,[9] or if X/O males are produced which carry this X, the males which appear show only traces of bristles and the abdominal integument is extremely etched. Some of these males show no traces of sexual terminalia, while the rest have very low fertility. If these males which are so strongly *bb* are crossed to $\widehat{XX}/Y^{-bb}$ females, males are obtained which have the same genetic constitution as their male parents as far as the sex chromosomes are concerned (Fig. 1). These backcross males, however, are no longer extreme *bb*.

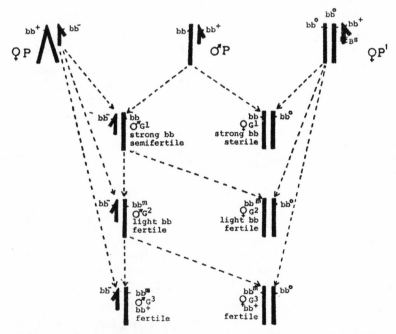

Fig. 1.—Scheme of experiments showing reversion of *bb* phenotype, while relevant sex chromosomes are the same. Only sex chromosomes are shown. The arrows show the origin of the chromosomes. Of the two arrows pointing to a certain genetic composition, one comes from a female and one from a male. Nonrelevant markers are omitted.

All the progeny have more or less the same phenotype. In parallel, males g^2 ty bb/Y^{-bb} which are extreme bb (males Gl in Fig. 1) were crosses to $sc^4sc^8/sc^4sc^8/Y^{B8}$ females (called P females in Fig. 1). Females g^2ty bb/sc^4sc^8 from this cross are very weak bb in phenotype and are now fertile. In this case also, the sex chromosome composition is identical to that of the corresponding Gl females, but a partial reversion of the bb phenotype toward the wild type has occurred. A much better reversion of the bb phenotype can be obtained for both the males and the females of the above-mentioned genotypes as follows: males g^2ty bb/Y^{-bb} (G2) are further crossed with females $\widehat{XX}/Y^{-bb}$ (females P of Fig. 1) of the original stock. Males g^2 ty bb/Y^{-bb} (G3 of Fig. 1) are no longer bb. If ty, which itself reduces the length of the bristles, and g^2 are removed from the original X chromosome, the same results are obtained.

When males g^2 ty bb/Y^{-bb} (G2) were crossed with sc^4 $sc^8/sc^4sc^8/Y^{B8}$ females, the great majority of progeny females of composition g^2 ty bb/sc^4sc^8 were also the wild phenotype.

To test whether the bb reversion was attributable to an increase in content of DNA complementary to ribosomal RNA, X/O males carrying the g^2 ty bb chromosome from a different origin were obtained. For one of these stocks, males g^2 ty tb/Y^+ (P males of Fig. 1) were crossed with $\widehat{XX}/O$ females, and g^2 ty bb/O were obtained which were extreme bb in phenotype.

In parallel, males g^2 ty bb/Y^{-bb} (Gl), which are extreme bb in phenotype, were crossed with the same $\widehat{XX}/O$ females used for the previous cross, and g^2 ty bb/O males were obtained which were very weakly bb in phenotype. Finally, males g^2 ty bb/Y^{-bb} (G3) were again crossed with $\widehat{XX}/O$ females, and g^2 ty bb/O males were obtained which were completely reverted to wild type from bb. The saturation levels in rRNA/DNA hybridization experiments, using DNA extracted from these three types of males, are reported in Figure 2. It is apparent that reversion of the $bobbed$ phenotype is paralleled by an increase in the content of DNA complementary to ribosomal RNA. These experiments show that the reversion of bb which occurs in passing from males Gl to males G2 and to males G3 (Fig. 1) is paralleled by an increase in rDNA, and this occurs in 100 per cent of the cases. I will call the reverted bb locus "magnified bobbed" (bb^m).

At this point it was important to know whether the rDNA produced in the process of reversion was firmly bound to the bb locus or not. If g^2 ty bb^m/Y^{-bb} males which are reverted (G3) are maintained with females which are $\widehat{XX}/Y^{-bb}$, they never show reappearance of the bb phenotype. The amount of rDNA associated with the X chromosome stabilizes around 0.27 per cent. Figure 3 shows the saturation level in a rRNA/DNA hybridization experiment of the DNA from g^2 ty bb^m/O males which were obtained by crossing $\widehat{XX}/O$ females with males g^2 ty bb^m/Y^{-bb}, which were kept for about ten generations with $\widehat{XX}/Y^{-bb}$ females.

I found, however, that if the reverted g^2 ty bb^m chromosome is put together with a partner Y chromosome carrying a bb^+ locus (in a male), then the bb^m locus, at least in part, returns to its original bb condition. Several experimental designs

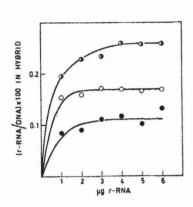
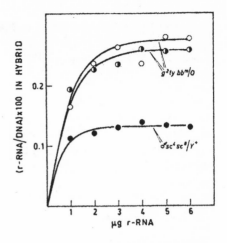

Fig. 2 (*left*).—Saturation levels in rRNA/DNA hybridization experiments with wild-type rRNA, H³-labeled, and DNA obtained from X/O males, obtained from the following crosses:

(●) ♀ $\hat{X}X$/O × g^2ty bb/Y⁺ (♂P of Fig. 1).

(○) ♀ $\hat{X}X$/O × g^2ty bb/Y⁻ᵇᵇ (♂G1 of Fig. 1).

(◑) ♀ $\hat{X}X$/O × g^2ty bb^m/Y⁻ᵇᵇ (♂G2 of Fig. 1).

The real saturation levels obtained in these experiments have been multiplied by 0.9 to take into account the lower DNA content of these individuals (which lack the Y chromosome). I expected the saturation level of g^2ty bb/O (●) males to be somewhat lower than the value observed, based on the intensity of the bb phenotype in these males. The explanation for this result can be a trivial one or involve one of the following possibilities: (*a*) certain tissues of these males have high levels of rDNA; (*b*) some genes of this particular locus are mutated.

Fig. 3 (*right*).—Saturation levels in rRNA/DNA hybridization experiments with wild-type rRNA, tritium-labeled, and DNA obtained from X/O males, obtained as follows:

(◑) $\hat{X}X$/O × g^2ty bb^m/Y⁻ᵇᵇ (this is a reproduction of the upper curve of Fig. 2).

(○) $\hat{X}X$/O × g^2ty bb^m/Y⁻ᵇᵇ (these males are obtained by keeping males G3 of Fig. 1 with females $\hat{X}X$/Y⁻ᵇᵇ for about ten generations).

(●) Males sc^4sc^8/Y⁺ used as control.

were followed to test this behavior, and one is presented in Figure 4. The results illustrated in Figure 5 show that at least a fraction of the bb loci that underwent "magnification" in only one generation can return to its original condition again in only one generation. In this case, however, the reversal is not always an all-or-none phenomenon. The classification, although somewhat arbitrary, is still feasible: the main criteria is abdominal etching. Since the intensity of the bb phenotype is an expression of the partial deficiency of rDNA,[1, 9] it is apparent that the rDNA which characterized the reversion of bb can be easily lost.

These results suggested that reversion of bb in passing from G1 males to G2 males and from these to G3 males (Fig. 1) might occur by accumulation of rDNA which is free from the chromosome. Were this true, the free rDNA would then be distributed at random between the gametes. Two experiments have been carried out to test this point: (1) g^2 ty bb^m/Y⁻ᵇᵇ (G3) males were crossed with sc^8bb^+/g^2 ty bb females. If "magnification" occurs by accumulation of free copies of rDNA and these are distributed at random between X and Y, one

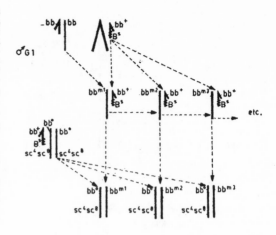

FIG. 4.—Scheme of the crosses used to keep the X chromosome carrying a bb^m locus with a Y chromosome carrying a bb^+ locus. The bb^m locus is tested, after several generations with a bb^+, by combining the chromosome which carries it with a sc^4sc^8X in a female (crosses indicated at the bottom of the figure). The sc^4sc^8 chromosome is also labeled bb^0 (no rDNA).

Substituting males P of Fig. 1 (g^2ty bb/Y^+) for males G1 (g^2ty bb/Y^{-bb}) in this scheme, one gets a control of the status of the "nonmagnified" bb after these crosses. The results of such controls are shown in Fig. 5.

should then find, from the above cross, males g^2 ty bb/Y^{-bb} which are no longer bb together with males of the same genetic composition which are extreme bb. The results of such a cross are shown in Table 1. A fraction of g^2 ty males that are no longer bb are found, while the majority are extreme bb. The bb^+ males, however, are all sterile and hence are X/O in chromosomal composition. They are the consequence of nondisjunction of the female X chromosomes. (2) $sc^4sc^8/$ g^2ty bb^m (G3) females, phenotypically bb^+, were crossed with sc^4sc^8/Y^{Bs} males. Were some rDNA carried along by the egg containing the sc^4sc^8 chromosome, one might expect to obtain females which are homozygous for the sc^4sc^8 chromosome (which carries no rDNA by itself)[2] and possibly having nucleoli free from the chromosomes, or small free nucleoli. No females of such type were obtained (Table 2).

From experiments (1) and (2), it is apparent that both the Y^{-bb} and sc^4sc^8 chromosomes are never associated with any free rDNA, even if they were partners of the g^2 ty bb^m chromosome.

Discussion.—This paper reports that, in terms of common experience, individuals of the same relevant chromosomal composition can have different phenotypes, depending upon the previous history of these elements. The X and

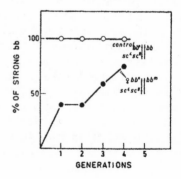

FIG. 5.—The figure shows, in successive generations, the percentage of strong bb females out of the total number of females of the following genetic constitution (sc^4sc^8/bb^{m1}; sc^4-sc^8/bb^{m2}; sc^4sc^8/bb^{m3}, etc.), obtained as indicated in Fig. 4. Controls are obtained as indicated in Fig. 4. The sc^4sc^8 chromosome is also labeled bb^0 (no rDNA).

164

TABLE 1. *Progeny test of the following cross:* $y^{31d}sc^9bb^+car\ cv/g^2ty\ bb \times g^2ty\ bb^m/Y^{-bb}$.

———Female Progeny———		———Male Progeny———	
Phenotype	N	Phenotype	N
Wild	223	y car cv	174
g^2ty	188	g^2ty^*	16
		g^2ty strong bb	32

* Sterile X/O males, product of nondisjunction of the females' X's or four-strand double crossing-over.

Y chromosomes of males G1, G2, and G3 of Figure 1 are the same, as are the X's of females G1, G2, and G3 of the same figure. Phenotypic differences between males and females G1, G2, and G3 cannot be accounted for by maternal effect, simple autosomal heterozygosis, nor by unequal crossing-over. To explain the difference in rDNA content between males G1 and males G2 in terms of unequal crossing-over, one would have to postulate an extraordinary frequency of this event between X and Y at the somatic level, and the favored element should always be the X chromosome. On the contrary, the frequency of crossing-over between X and Y in males was estimated to range from 2×10^{-4} to 8×10^{-4}.[10]

I previously showed that the Y^{-bb} chromosome[11] carries about 100 genes for ribosomal RNA which are not operative. The possibility of direct involvement of these or other genes of the Y in the phenotypic differences between males G1, G2, and G3 was discounted by producing a series of X/O males through crossing $\widehat{XX}$/O females with males P, G1, and G2, respectively, of Figure 1. The males obtained from these crosses are, respectively, strong bb, light bb, and wild type. The observed phenotypic differences between females G1, G2, and G3 lead also to the same conclusion.

After these considerations I looked for template variation and found that males G1 have a lower content of genes for rRNA than males G2, and these have a lower content than males G3 (Fig. 2). The number of genes does not augment any further if males G3 are permanently kept with females P (Fig. 3). The conclusion then is that reversion of bb in males G2 and G3 occurs by "magnification" of genes specific for rRNA. "Magnification," furthermore, cannot occur at the level of all somatic cells. If the combination X^{bb}/Y^{-bb} were the only condition to acquire "magnification," then all body cells of males G1 should have a normal, not a bb phenotype: the phenomenon would be undetectable in the approach used in the present investigation. Somatic cells instead appear no longer bb

TABLE 2. *Progeny test of the following cross:* $y\ sc^4sc^9car\ cv/g^2ty\ bb^m(G3) \times y\ sc^4sc^9car\ cv/Y^{B^s}$.

———Female Progeny———		———Male Progeny———	
Phenotype	N	Phenotype	N
$B^s*‡$	81	y car cv $B^s†‡$	171
y car cv $B^s*‡$	58	$g^2ty\ B^s‡$	126
Wild†	184	g^2ty^*	161
Light bb†	109	Wild*	13
Strong bb†	5		

* Nondisjunctional products.
† Some might come from nondisjunction in both males and females.
‡ Few, nonrelevant, double crossovers have been included in the main group.

when the X^{bb} chromosome comes from males G1. Since a bb phenotype is the expression of the number of genes for rRNA, it seems quite reasonable to conclude that the "magnification" of rDNA occurred in the germ line cells of males G1. From the observation that all cells of males G2 show a reversion of the bb character, and remembering that bb is a cell-autonomous character,[12] we can also conclude that the rDNA produced in the "magnification" process is maintained in all progeny cells; it is hence self-duplicating. A possible explanation of the phenomenon would assume a strong selection at the level of the original X^{bb}/Y^{-bb} spermatogonia; only those in which unequal crossing-over has led to an increased number of genes for rRNA would generate active sperm. On this assumption, however, the bb^m locus should, afterward, behave in a completely normal fashion. This kind of rDNA cannot, however, be considered perfectly integrated into the chromosome. Were it so, one would expect the bb^m locus to be as stable as any bb locus and hence to produce partial deletions with a frequency lower than 0.6 per cent. On the contrary, if a bb^+ locus is a partner of bb^m, bb^m shows a high frequency of reversion to the original bb condition. The frequencies of revertants shown in Figure 5 are possibly underestimated, since a fraction of strong bb individuals die before hatching. A control to this phenomenon is the daily experience of the stability of the bb^+ locus. However, a direct measure of it is illustrated in Figure 4. Were the bb^+ locus of the Y^{B^s} chromosome unstable as the bb^m locus, one should find bb mutations associated to this chromosome as frequently as observed in its partner. This was never observed to be the case in this and similar controls involving X^{bb^+}/X^{bb^m} females. From the behavior of bb^m in the presence of a bb^+ partner, one could visualize the bb^m locus as constituted of two parts: one stably integrated within the chromosome (the original bb locus), and the other relatively free (diluted 30–50% per generation). On these bases one could expect to find some of these hypothetical free copies to be inherited along with the partner of the bb^m locus after segregation. This was found to be not so when the partners were the sc^4sc^8 and the Y^{-bb} chromosomes (Tables 1 and 2).

One could imagine that the amount of rDNA measured in the adults of *Drosophila* is itself a product of a "tissue-directed magnification" of a few chromosomally integrated genes for rRNA; bb mutations could hence be conceived as alteration of the normal "magnification" mechanism. The data presented here, in this case, could involve alteration of this kind of a mechanism. Data pointing to chromosomal integration of many copies of genes for rRNA are, however, rapidly accumulating.[3, 4, 7]

Summary.—The data presented here show that sudden reversion of bb (bb to bb^m) can occur in certain circumstances and that its molecular explanation is due to an accumulation of rDNA; bb^m, however, can return with high frequency to its original bb condition if matched with a normal bb^+ locus. A direct test to see whether the return to the original bb condition is paralleled by loss of rDNA has not yet been made. Considerable data[1, 9] on the parallelism between bb phenotype and rDNA content suggest, however, that this is probably the case. Given these observations, two working hypotheses can be considered to explain the observed phenomenon. (1) Certain chromosomes can undergo selective increase and selective loss of rDNA. While selective increase could be easily accounted for by spermatogonial selection of bb^+ cells originated after unequal crossing-over or intrachromosomal exchange within the bb loci, it is difficult at present to conceive how directed losses of rDNA in a particular one of the chromosomes of a spermatogonia could have selective advantage. (2) Mechanisms exist which allow independent duplication of specific chromosomal sections. The genome fractions, thus duplicated, are capable of only loose integration within the chromosome. Examples of differential duplication of specific chromosomal sections are known,[13-16] as are examples of instability of certain genetic situations (see, for example, Brink et al.[17]). The molecular events leading to "magnification" of rDNA in our case are obscure. The phenomenon, however, is not restricted to the particular case presented here but always occurs, even if the efficiency can variate, in males of strong bb phenotype (to be published elsewhere). The possibility that this phenomenon is not restricted to the genes for rRNA must be entertained.

I thank Professors T. Sonneborn, J. Schultz, G. Magni, Dr. C. Auerbach, Dr. R. C. von Borstel, and, in particular, Professor S. Spiegelman for reviewing this manuscript and for critical discussion. The continuous skillful assistance of Mr. G. Scala was greatly appreciated.

* Work carried out under the Association Euratom-CNR-CNEN, contract no. 012-61-12 BIAI.
[1] Ritossa, F. M., K. C. Atwood, and S. Spiegelman, *Genetics*, **54**, 819 (1966).
[2] Ritossa, F. M., and S. Spiegelman, these PROCEEDINGS, **53**, 737 (1965).
[3] Birnsteil, M. L., H. Wallace, J. L. Sirlin, and M. Fischberg, *Natl. Cancer Inst. Monograph*, **23**, 431 (1966).
[4] Ritossa, F. M., K. C. Atwood, D. L. Lindsley, and S. Spiegelman, *Natl. Cancer Inst. Monograph*, **23**, 449 (1966).
[5] Brown, D. D., in *Current Topics in Developmental Biology*, ed. A. Monroy and A. Moscona (New York: Academic Press, 1967), vol. 2, p. 47.
[6] Spiegelman, S., and W. F. DeLorenzo, thesé PROCEEDINGS, **38**, 583 (1952).
[7] Schalet, A., *Genetics*, **56**, 587 (1967).
[8] Gillespie, D., and S. Spiegelman, *J. Mol. Biol.*, **12**, 829 (1965).
[9] Ritossa, F. M., these PROCEEDINGS, **59**, 1124 (1968).
[10] Cooper, K. H., *Chromosoma*, **10**, 535 (1959).
[11] Schultz, J., in C. B. Bridges, and K. S. Brehme, *Carnegie Institution of Washington Publ.* **552** (1944), p. 233.
[12] Brosseau, G. E., *Drosophila Information Service*, **33**, 122 (1959).
[13] Ficq, A., and C. Pavan, *Nature*, **180**, 983 (1957).
[14] Rudkin, G. T., and S. L. Corlette, these PROCEEDINGS, **43**, 964 (1957).
[15] Miller, O. L., *J. Cell Biol.*, **23**, 60A (1964).
[16] Peacock, W. J., *Natl. Cancer Inst. Monograph*, **18**, 101 (1965).
[17] Brink, R. A., E. D. Styles, and J. D. Axtell, *Science*, **159**, 161 (1968).

Localization of Deoxyribonucleic Acid Complementary to Ribosomal Ribonucleic Acid and Preribosomal Ribonucleic Acid in the Nucleolus of Rat Liver*

WILLIAM J. STEELE

In recent years, there has been an impressive accumulation of evidence derived from a correlation of cytochemical and genetic information with data obtained from molecular hybridization in support of the hypothesis that the nucleolar organizer is the site of synthesis of ribosomal RNA (1–4). Early attempts to show the localization of DNA complementary to ribosomal RNA in isolated nucleoli of pea seedlings (5) and HeLa cells (6) have met with variable success, presumably because of technical difficulties involved in the isolation of highly purified nucleoli.

A number of studies have shown that the nucleolus apparently plays a key role in the synthesis (7–13), methylation (14, 15), and interconversion (8, 9, 12, 16–19) of the rapidly sedimenting nucleolar RNAs. Nevertheless, the relationship between nucleolar preribosomal RNAs and cytoplasmic ribosomal 18 S RNA and 28 S RNA has not been satisfactorily defined. Generally, it has been assumed that a single rapidly sedimenting intermediate is cleaved to produce an 18 S RNA, which is rapidly expelled from the nucleolus and nucleus, and a 28 S RNA or 35 S RNA, which is detained in the nucleolus while conversion to a ribonucleoprotein takes place (8, 9, 12). On the other hand, kinetic analysis of events occurring in isolated nucleoli of liver of normal (12, 17) and thioacetamide-treated (20) rats and in the nuclear ribonucleoprotein network of the nucleus (17) have suggested that ribosomal 18 S RNA and 28 S RNA may be

* These studies were supported in part by Grants BUCM 30-2658 and CA-08182 from the United States Public Health Service and by grants from The National Science Foundation, The American Cancer Society and the Jane Coffin Childs Fund. A preliminary report of this work has appeared (STEELE, W. J., *Fed. Proc.*, **27**, 805 (1968)).

synthesized at different sites in the nucleus. Because of the relative ease with which highly purified nucleoli may be obtained from isolated nuclei of rat liver (21), it seemed possible to investigate these questions by the technique of RNA-DNA hybridization.

In this study it was found that the nucleolar DNA fraction of rat liver contained the bulk of DNA sequences complementary to both ribosomal 18 S RNA and 28 S RNA, and to nucleolar 28 S RNA, 35 S RNA, 45 S RNA, and 55 S RNA. A major part of these sequences was associated with a heavy DNA satellite component, suggesting that the multiple ribosomal RNA cistrons of the nucleolus may be contiguous. In addition the conversion of preribosomal nucleolar RNAs to ribosomal RNAs involved a stepwise process of chain shortening.

MATERIALS AND METHODS

Albino male rats weighing 150 to 175 g were obtained from the Cheek-Jones Company, Houston, Texas. Other materials were obtained from the following sources: orotic acid-5-^{3}H from Nuclear-Chicago (5 C per mmole) and Schwarz BioResearch (13.8 C per mmole); α-amylase, lysozyme, RNase A, and RNase T_1 from Worthington; Pronase from Calbiochem; and sucrose, medium coarse granular and liquid concentrate (68%, w/w) from local suppliers (22). The 0.34 M and 0.88 M sucrose solutions were passed through Millipore filters (HA, 0.45-μ). The SSC buffer, which contains 0.15 M NaCl and 0.015 M sodium citrate, was prepared at two concentrations (1 $\times$ SSC and 24 $\times$ SSC), passed through Millipore filters, and diluted as needed.

Isolation of Nucleoli—For the preparation of isolated nucleoli of liver tissue, isolated nuclei were obtained first by a modification of the method of Chauveau, Moulé, and Rouiller (23) as previously described (22), with the exception that the tissue was homogenized in a jacketed, glass-Teflon homogenizer (Glenco, Houston, Texas) which was cooled by circulating ice water. The nuclear pellets were suspended in 0.34 M sucrose (nuclei from 1.5 g of tissue per ml) by homogenization with a glass-Teflon homogenizer (6 to 9 $\times$ 10^{-3} inch clearance) with the pestle rotating at 800 rpm. An aliquot (30 to 50 ml) of the nuclear suspension was poured without foaming into a rosett cooling cell (Heat Systems Company, Melville, New York, model 50) immersed in an ice bath. The suspension was treated (Branson Sonifier, model S-125) generally at full power (9 to 10 amp) for 10-sec periods with a solid, highly polished stephorn converter immersed to a depth of 1 inch. Smaller volumes of suspension (30 ml) and lower power settings were used to shorten the treatment time, which helped to preserve the integrity of nucleoli. The progress of nuclear disruption was monitored by phase contrast microscopy until intact nuclei were absent from

169

at least 10 low power fields (usually 30 to 40 sec for 50 ml). The sonic extract was aspirated with a cold syringe (50 ml), transferred to centrifuge tubes, and underlayered with cold 0.66 or 0.88 M sucrose (0.5 volume). Isolated nucleoli were precipitated in pellet form by centrifugation at 6000 × *g* for 10 min (Sorvall RC-3, HG-4 rotor) at 0°. The nucleolar fraction was essentially free from contamination by isolated nuclei (1 per 4000 nucleoli). The number of isolated nuclei and nucleoli in the preparation was determined as previously described (10).

Isolation of DNA—DNA was extracted from nuclear and nucleolar pellets with a solution containing 5% sodium 4-aminosalicylate (24), 1% sodium dodecyl sulfate, and 0.1 M EDTA, pH 8, by homogenization at 1,000 rpm with a loose fitting pestle of a glass-Teflon homogenizer. The volume of solution was adjusted to maintain a DNA concentration of at least 50 μg per ml. An equal volume of phenol-cresol mixture containing 0.1% 8-hydroxyquinoline (25) was added and the mixture was homogenized as before. The mixture was shaken at room temperature for 5 min and centrifuged at 30,000 × *g* for 10 min. The aqueous layer was removed with a wide bore pipette and extracted with fresh phenol-cresol mixture as before. The nucleic acids were precipitated from the aqueous layer with ethanol, sedimented, and dissolved in 0.1 M NaCl. The precipitation step was repeated and the pellet was dissolved in 1 × SSC.

An alternative method was used for the sequential extraction of DNA and RNA from isolated nucleoli. Nucleolar pellets were homogenized initially with 2 M NaCl (0.2 ml per g of original tissue) at 2°. The insoluble nucleolar residue fraction, which contained the bulk of the nucleolar RNA (22), was precipitated as a pellet by centrifugation at 30,000 × *g* for 20 min and extracted again with fresh 2 M NaCl. The combined supernatant solutions, which contained the bulk of the nucleolar DNA, were treated with cold ethanol (2 volumes) to precipitate crude DNA and centrifuged. The pellet was dissolved in the sodium aminosalicylate-sodium dodecyl sulfate-EDTA medium described in the first procedure and then deproteinized as before. The salt-insoluble residue of the nucleolus was treated for extraction of RNA according to methods described below.

Carrier DNA for RNA-DNA hybrid recovery studies was prepared from *Bacillus megaterium* spheroplasts (26) according to the first procedure described above.

Purification of DNA—All crude DNA samples were incubated in 1 × SSC with a mixture of α-amylase (50 μg per ml), RNase A (150 μg per ml), and RNase T₁ (10 units per ml) for 1 hour at 37°. Pronase was added at a level of 100 μg per ml, and the incubation was continued for an additional 2 hours at 37° (27). The mixture was diluted by the addition of an equal volume of

1% sodium dodecyl sulfate solution (10) and deproteinized as described above. Aliquots of DNA for base composition analysis were purified further by centrifugation on CsCl gradients (28). The DNA was recovered from CsCl solutions by the addition of water (2 volumes), followed by precipitation with an equal volume of ethanol.

Preparation of Labeled Ribosomal RNA—One rat was given an intraperitoneal injection of orotic acid-5-^{3}H (5 to 10 mC), food was withdrawn, and the rat was killed 20 hours later. The liver was perfused with cold 0.25 M sucrose *in situ* and removed. The C-ribosome fraction was prepared and incubated for 15 min at 37° according to the methods described by Wettstein, Staehelin, and Noll (29). The resulting preparation of mono- and disomes was treated with sodium pyrophosphate (30) and fractionated into 30 S and 50 S ribosomal subunits on sucrose density gradients (30). Ribosomal RNA was isolated from the fractionated subunits by extraction with 0.5% sodium dodecyl sulfate solution (10) and deproteinization with phenol-cresol mixture as described above. Aliquots of the RNA fractions were centrifuged on linear sucrose gradients (10), and the middle portions of the 18 S RNA and 28 S RNA peaks were separated with the aid of a model D density gradient fractionator (Instrumentation Specialties Company, Lincoln, Nebraska) for use in this study. The specific radioactivities of the two rRNAs were 2000 to 3500 cpm per μg.

Preparation of Labeled, Rapidly Sedimenting Nucleolar RNAs—The rats (12 to 24) were given intravenous injections of 250 μC of orotic-acid-^{3}H and killed 40 min later. Isolated nuclei and nucleoli were prepared from liver tissue as described above. Nucleolar RNA was extracted by one of the following methods. Isolated nucleoli were extracted first with 2 M NaCl to remove DNA as described above, and then the residue was treated with 1% sodium dodecyl sulfate solution and deproteinized (31). Nucleolar RNA was also extracted selectively from isolated nucleoli with 0.3% sodium dodecyl sulfate solution and deproteinized with phenol-cresol mixture as previously described (31). Purified RNA was fractionated on linear sucrose density gradients (1.5 mg per 28-ml tube or 0.5 mg per 16 ml) and the peaks corresponding to 28 S RNA, 35 S RNA, 45 S RNA, and 55 S RNA were precipitated with ethanol containing 2% potassium acetate. Individual RNA fractions were centrifuged on new sucrose gradients and fractionated as before until single, symmetrical peaks of optical density were obtained for each of the rapidly sedimenting nucleolar RNA fractions. The specific radioactivities of the RNA fractions were 2000 to 3000 cpm per μg.

Hybridization—The procedure described by Gillespie and Spiegelman (27) was used for all studies with immobilized DNA.

171

Hybridization was carried out in 6 × SSC at 67° for 16 hours. The 6 × SSC buffer solution was also used for both immobilization of the denatured DNA and washing procedures. Liquid hybridization of DNA fractions derived from CsCl gradients was performed as follows. Each fraction was dialyzed against frequent changes of 0.1 × SSC for 2 days at 4° to remove CsCl and then diluted by the addition of the same buffer to 1.0 ml. *B. megaterium* DNA (25 μg) was added to each fraction. The DNA mixture was denatured by heating at 100° for 10 min and then chilled rapidly in an ice bath. The concentration of SSC was adjusted to 2 × SSC before labeled RNA was added. The fractions were incubated at 67° for 2 hours, chilled in ice water, diluted by the addition of 6 × SSC (5 volumes), and passed through a membrane filter (Bac-T-Flex B-6, Schleicher and Schuell, Keene, New Hampshire) to trap the hybrid (32). The filter discs were washed, treated with RNase, and assayed for radioactivity according to the procedures described by Gillespie and Spiegelman (27).

Base Composition—Aliquots of DNA (0.5 mg) were digested with 12 N perchloric acid (0.02 ml) for 1 hour at 100° (33), diluted by the addition of water (2 ml), and centrifuged to remove carbonaceous by-products. The supernatant solution containing DNA bases was added directly to a Dowex 50-X8, hydrogen form, column (0.5 × 30 cm), and fractions (3 ml) were collected from the beginning. The bases were eluted with a linear gradient of HCl (water to 4 N HCl). Three analytical columns and one blank column were supplied from a common reservoir at the same flow rate (0.3 ml per min). Fractions were monitored for optical density at 260 mμ. The individual peaks were pooled into volumetric containers and diluted by the addition of water to the same volume (25 ml). An equivalent number of fractions were pooled from the blank column for each of the four peaks and diluted as before. The optical density was determined for each base at the appropriate wave length against a matching blank. Whereas the millimolar extinction coefficient of adenine was 11.9 at 262 mμ under these conditions, extinction coefficients for the other bases were the same as published values at the wave length of maximum absorption (34). A standard mixture of pure DNA bases was chromatographed at the same time as many nucleolar DNA samples to monitor the recovery of bases from the column.

Aliquots of RNA (0.5 to 1 mg) were hydrolyzed with 0.3 N KOH for 18 hours at 37° and nucleotides were fractionated according to the methods described previously (10).

Analytical Methods—RNA and DNA were determined according to the methods previously described (17) on perchloric acid extracts of tissue subfractions.

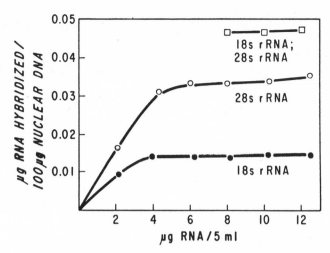

Fig. 1. Saturation curves for annealing ³H-ribosomal 18 S RNA and 28 S RNA with nuclear DNA of rat liver. Membrane filters containing 100 μg of denatured DNA were immersed in 5 ml of 6 × SSC containing increasing amounts of RNA and heated at 67° for 16 hours. Hybrids were purified as described (27). Tests for additivity were performed by annealing first with saturating amounts of one RNA, washing with 6 × SSC, annealing with saturating amounts of a second RNA, and then purifying the hybrids as before (27).

RESULTS

Nuclear DNA Complementary to rRNAs—Fig. 1 shows the results of a series of studies in which increasing amounts of homologous tritium-labeled 18 S rRNA or 28 S rRNA were annealed with a constant amount of DNA of isolated nuclei of rat liver tissue. The plateaus of the curves indicate that the fractions of nuclear DNA complementary to 18 S rRNA and 28 S rRNA were about 1.4×10^{-4} and 3.2×10^{-4}, respectively. The ratio of these values corresponds closely to the ratio of the average molecular weights of the rRNAs (35) and to the ratio of amounts of the rRNAs found in isolated ribosomes (29). In view of the precautions taken to remove mRNA from rRNA, it is possible that the slight increment in the hybridization of 28 S rRNA in the plateau region of the curve (Fig. 1) may have been due to contamination of 28 S rRNA with a small amount of 18 S rRNA.

A two-step hybridization technique (2) was required to ascertain whether distinct sequences of nuclear DNA were complementary to each of the two rRNAs, because initial studies showed that there was a significant amount of competition between 18 S rRNA and 28 S rRNA for nuclear DNA sequences when the two were present at saturating amounts (36). The data (Fig. 1, *top curve*) show that the fraction of nuclear DNA complementary to the combined rRNA sequences was about 4.7×10^{-4}. This fraction was nearly identical with the sum of the fractions found for the individual rRNAs. Similar results were obtained irrespective of the sequence in which the rRNAs were annealed with the DNA.

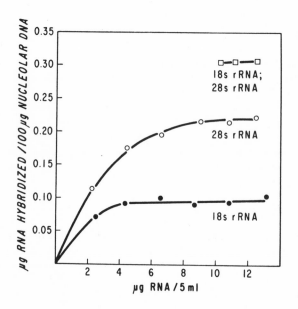

Fig. 2. Saturation curves for annealing ^{3}H-ribosomal 18 S RNA and 28 S RNA with the DNA fraction of isolated nucleoli of rat liver. The procedures were the same as those described in the legend for Fig. 1, except that 50 µg of denatured nucleolar DNA were used.

Nucleolar DNA Complementary to rRNAs—Fig. 2 shows the results of a series of studies in which increasing amounts of homologous, tritium-labeled 18 S rRNA or 28 S rRNA were annealed with a constant amount of DNA of isolated nucleoli of rat liver. The plateaus of the curves indicate that the fractions of nucleolar DNA complementary to 18 S rRNA and 28 S rRNA were about 0.97×10^{-3} and 2.2×10^{-3}, respectively. In addition, the results of studies in which the two rRNAs were annealed sequentially with nucleolar DNA (Fig. 2, *top curve*) showed that the fraction of nucleolar DNA complementary to the combined rRNA sequences was about 3.1×10^{-3}. This fraction was nearly identical with the sum of the fractions found for the individual rRNAs, indicating that distinct sequences of nucleolar DNA were complementary to each of the two rRNAs. Moreover, the evidence that both 18 S rRNA and 28 S rRNA produced about 7 times as much hybrid with nucleolar DNA as with a similar amount of nuclear DNA supports the concept that the nucleolus is the site of synthesis of the two rRNAs.

To investigate the recovery of rRNA cistrons in isolated nucleoli, the amount of DNA and its complementarity to rRNA was determined for each of the nuclear subfractions obtained from the preparation of isolated nucleoli. The data presented in Table I show that 28 to 30% of the total rRNA cistrons were recovered in the nucleolar DNA fraction (pellet). This indicated that at least 75% of the rRNA cistrons were confined in the isolated nucleolus, since only 35 to 40% of the total nucleoli were recovered in the pellet.

TABLE I

Distribution of DNA and rRNA cistrons in nuclear subfractions
of rat liver

In these studies the conditions for the preparation of isolated nucleoli, which are discussed under "Materials and Methods," were optimized to increase the yield and preserve the integrity of nucleoli at the possible expense of a slight contamination by isolated nuclei. Saturation curves for rRNAs were obtained according to the methods described in the legend for Fig. 1. The percentage hybridization values for rRNAs are averages of triplicate determinations from three separate experiments. The values in parentheses are percentages of the total 18 S rRNA and 28 S rRNA cistrons of the nucleus.

Fraction	Distribution of DNA	DNA complementary to	
		18 S rRNA	28 S rRNA
	%	%	%
Whole sonic extract of nuclei	100.0	0.010 (100)	0.026 (100)
0.34 M sucrose layer	96.5	0.007 (68)	0.017 (63)
0.88 M sucrose layer	0.3	0.085 (3)	0.19 (2)
Pellet (nucleoli)	3.0	0.10 (30)	0.24 (28)

175

Sedimentation Analysis of Nucleolar DNA Fraction—DNA isolated from nucleoli and fractionated by preparative centrifugation in CsCl (Fig. 3a) contained a main peak that had a buoyant density of 1.700 (analytical ultracentrifuge, bacteriophage SP8 DNA marker), a lighter satellite component, and a heavier satellite component. Centrifugation of fractions removed from the main peak (Fraction 15) and the heavy satellite component (Fraction 18) in fresh CsCl (Fig. 3, b and c, respectively) showed that the heavy satellite had a buoyant density of 1.708 (*C. perfringens* DNA marker, not shown) and contained about 8% of the total nucleolar DNA. Similar studies showed that the light satellite component had a buoyant density of about 1.692 and contained about 4% of the total DNA.

Thermal denaturation studies on aliquots of nucleolar DNA and nuclear DNA showed that the profiles of the nucleolar DNA fraction were indistinguishable from profiles of whole nuclear DNA at several concentrations of SSC. Increases in hyperchromicity were 36 to 38%, and T_m values were 72–73° in 0.1 × SSC for both the nuclear and the nucleolar DNA fractions.

Nucleolar DNA Subfractions Complementary to rRNAs—To determine which nucleolar DNA components were complementary to rRNA, hybridization studies were performed with nucleolar DNA fractions collected from CsCl gradients (Fig. 3a) and saturating amounts of homologous, tritium-labeled rRNAs. Representative profiles (Fig. 4) show that DNA complementary to 18 S rRNA was distributed in two peaks; the smaller of the two peaks was associated with the main DNA peak and the larger of the two peaks was associated with the heavy satellite shoulder. DNA complementary to 28 S rRNA was associated predominately with the main DNA peak and to a lesser extent with the heavy satellite shoulder. The percentage hybridization of 18 S rRNA and 28 S rRNA (Fig. 4) attained peak values of 1.0 and 2.0, respectively, in a region of the CsCl gradient of higher buoyant density than that of the heavy satellite. This suggests that much higher values may be obtained for the heavy DNA satellite by further fractionation.

Saturation Plateaus of Rapidly Sedimenting Nucleolar RNAs—To investigate the relationship between cytoplasmic rRNAs and postulated preribosomal RNAs of the nucleolus, homologous, tritium-labeled RNA was prepared from the nucleolar residue fraction (Fig. 5) and separated into single, symmetrical peaks with approximate sedimentation coefficients of 28 S, 35 S, 45 S, and 55 S by repeated fractionation on sucrose density gradients. Fig. 6 shows the results of a series of studies in which increasing concentrations of individual rapidly sedimenting nucleolar RNA fractions were annealed with a constant amount of nucleolar DNA fraction. The plateaus of the curves indicate that the fractions of nucleolar DNA complementary to nucleolar 28 S

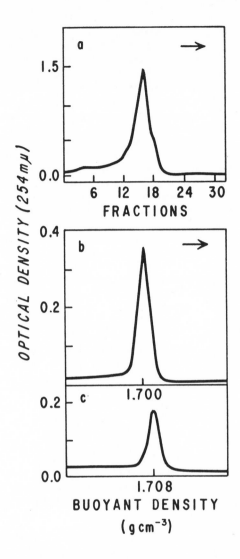

FIG. 3. CsCl density gradient profiles of nucleolar DNA fractions of rat liver. Purified nucleolar DNA (100 to 150 μg/0.7 ml) was layered over 3 ml of CsCl (1.80 g per cm³) and centrifuged for 18 hours at 25° at 39,000 rpm (Spinco SW 39 rotor). Optical density (254 mμ) was monitored and 6-drop fractions were collected with the aid of an ISCO density gradient fractionator (a). Fractions from the main peak (b) and the heavy satellite shoulder (c) were dialyzed and centrifuged on fresh CsCl solution, with *Clostridium perfringens* DNA marker (not shown). The *arrows* indicate the direction of sedimentation.

RNA, 35 S RNA, 45 S RNA, and 55 S RNA were about 4.4 × 10^{-3}, 4.4 × 10^{-3}, 5.2 × 10^{-3} and 5.3 × 10^{-3}, respectively.

Base Sequence Relationships among rRNAs and Nucleolar RNAs—Similarities between the primary sequences of the two rRNAs and the rapidly sedimenting nucleolar RNAs were also investigated by the two-step hybridization procedure (2), because of the spurious competition previously noted between mammalian rRNAs with different primary sequences (36). Nucleolar DNA was annealed with saturating amounts of 18 S rRNA or 28 S rRNA in the first step and then with saturating amounts of tritium-labeled rRNAs or nucleolar RNAs in the second step. The data presented in Table II show that saturation values were obtained for both 18 S rRNA and 28 S rRNA, irrespective of the sequence in which they were annealed with the DNA. The competition studies between rRNAs and nucleolar RNAs showed that nucleolar 28 S RNA, 35 S RNA, and 45 S RNA shared about 35, 34, and 30%, respectively, of their primary sequences in common with 18 S rRNA, and about 70, 75, and 50%, respectively, of their primary sequences in common with 28 S rRNA. In other studies it was shown that nucleolar 28 S RNA and 35 S RNA shared about 85% of their primary sequences in common with nucleolar 45 S RNA.

Base Composition of rRNAs and Nucleolar RNAs—Table III presents the base composition of ribosomal 18 S RNA and 28 S RNA and nucleolar 28 S RNA, 35 S RNA, and 45 S RNA of rat liver as determined by optical density. As noted previously (8, 25), the base composition of 18 S rRNA was different from the base composition of 28 S rRNA, mainly in the higher content of adenine and uracil and lower content of cytosine and guanine of 18 S rRNA compared to 28 S rRNA. The analyses for the base composition of the rapidly sedimenting nucleolar RNAs showed that the conversion of nucleolar 45 S RNA to nucleolar 28 S RNA (12, 17, 19) was accompanied by a progressive increase in the content of adenine and decrease in the content of uracil of the respective RNAs. With the exception of a slightly higher content of guanine for nucleolar 35 S RNA compared to nucleolar 45 S RNA, no apparent trends were found in the amounts of either cytosine or guanine. The base composition of 28 S rRNA was also different from the base composition of nucleolar 28 S RNA, mainly in the higher content of adenine and lower content of cytosine and guanine of 28 S rRNA compared to nucleolar 28 S RNA.

Nucleolar DNA Fractions Complementary to Preribosomal RNAs—To further characterize the nucleolar DNA sequences common to the rRNAs and the preribosomal RNAs of the nucleolus, hybridization studies were performed with nucleolar DNA fractions collected from CsCl gradients and saturating amounts of tritium-labeled nucleolar 28 S RNA, 35 S RNA, and 45 S RNA. The representative profiles in Fig. 7 show that

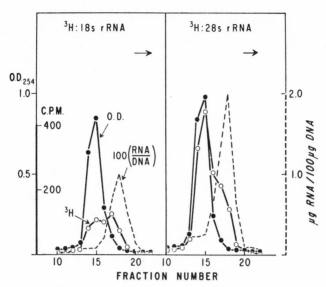

FIG. 4. Buoyant density distribution of nucleolar DNA complementary to ³H-ribosomal 18 S RNA and 28 S RNA. The procedures were the same as those described in the legend for Fig. 3, except that optical density was determined again after dialysis of the fractions against 0.1 × SSC. *B. megaterium* DNA (25 μg) was added to each fraction and denatured by heating. Each fraction was annealed with RNA (5 μg) and the hybrids were purified as described in "Materials and Methods."

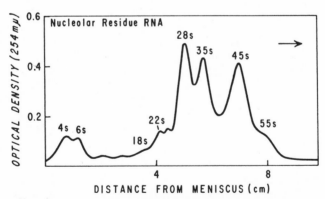

FIG. 5. Sedimentation profile of the RNA of the nucleolar residue fraction of rat liver. Purified RNA (0.15 mg) was layered over a linear 5% to 40% (w/w) sucrose density gradient (10) and centrifuged for 16 hours at 4° at 25,000 rpm (Spinco SW 25.3 rotor). The *arrow* indicates the direction of sedimentation. The *numbers* above the *peaks* indicate the approximate sedimentation coefficients.

DNA complementary to nucleolar 28 S RNA, 35 S RNA, and 45 S RNA was distributed in a relatively broad zone. The peak fractions were associated with DNA fractions of buoyant densities either between the main DNA peak and the heavy satellite shoulder or almost coincidental with the heavy satellite shoulder, as seen for nucleolar 45 S RNA. As noted previously for the rRNAs (Fig. 4), the percentage hybridization of nucleolar 28 S RNA, 35 S RNA, and 45 S RNA (Fig. 7) attained peak values of about 2.0 in a region of the CsCl gradient of higher buoyant density than that of the heavy satellite. This was presumably due to the inadequate separation of heavy satellite from the main DNA peak obtained by preparative centrifugation in CsCl.

Base Composition of DNA—Table IV presents the base composition of the nucleolar and nuclear DNA fractions of rat liver and *B. megaterium* DNA as determined by optical density. The base composition of nucleolar DNA was different from the base composition of nuclear DNA, mainly in the higher content of adenine and lower content of thymine of nucleolar DNA com-

TABLE II

Competition between ribosomal RNAs and nucleolar RNAs for DNA of isolated nucleoli of rat liver

The table presents percentages of the established saturation values for the RNAs that were used in the second step of annealing with nucleolar DNA. About 50 to 60 μg of denatured DNA were fixed to each disc. The details of the two-step hybridization procedure are described under "Materials and Methods." In preliminary studies, not shown in the table, it was found that tritium-labeled hybrids formed in the first step of annealing were completely stable in the presence (or absence) of either rRNAs or nucleolar rRNAs used in the second step of annealing. The values are averages of four determinations.

RNA fraction, first and second steps	Saturation for RNA used in second step
	%
18 S rRNA	
28 S rRNA	96
28 S nucleolar RNA	65
35 S nucleolar RNA	66
45 S nucleolar RNA	70
28 S rRNA	
18 S rRNA	95
28 S nucleolar RNA	30
35 S nucleolar RNA	25
45 S nucleolar RNA	50

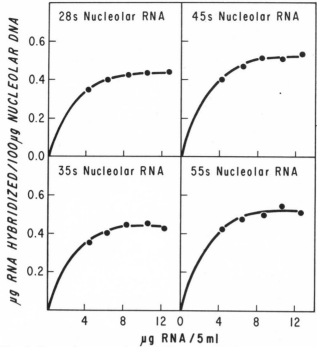

FIG. 6. Saturation curves for annealing ³H-nucleolar 28 S RNA, 35 S RNA, 45 S RNA, and 55 S RNA with the DNA fraction of isolated nucleoli of rat liver. The procedures were the same as those described in the legend for Fig. 1, except that 50 μg of denatured nucleolar DNA were used.

Base compositions of ribosomal RNAs and nucleolar RNAs of rat liver

The table presents the composition, in moles per 100 moles, of the given nucleotide as determined by ultraviolet analysis of nucleotides fractionated by ion exchange chromatography. The values are averages of three to six analyses. The sedimentation coefficients shown in Fig. 5 and referred to in the text are approximate modal values for the given peak.

RNA fraction	Base composition				Base ratio, (Ade + Ura): (Gua + Cyt)
	Ade	Ura	Gua	Cyt	
	moles/100 moles				
Ribosomal					
18 S	20.3	20.5	34.0	25.2	0.69
28 S	16.3	16.8	37.5	29.4	0.49
Nucleolar					
28 S	14.7	16.6	38.2	30.4	0.46
35 S	13.5	17.0	38.9	30.6	0.44
45 S	12.9	19.3	37.2	30.6	0.47

TABLE IV

Base compositions of nucleolar DNA and nuclear DNA of rat liver

The table presents the content, in moles per 100 moles, of the given base as determined by ultraviolet analysis of the bases fractionated by ion exchange chromatography. The values are averages of four to six analyses.

DNA	Base composition				Base ratio, (Ade + Thy): (Gua + Cyt)
	Ade	Thy	Gua	Cyt	
	moles/100 moles				
Nucleolar, rat liver.....	29.9	27.1	21.6	21.5	1.32
Nuclear, rat liver.......	28.8	28.7	21.2	21.3	1.35
B. megaterium..........	30.2	30.8	19.0	19.9	1.57

pared to nuclear DNA. The base ratio (adenine + thymine to guanine + cytosine) of nucleolar DNA was nearly the same as that for nuclear DNA, because the differences were only in the amounts of adenine and thymine. The base composition of *B. megaterium* DNA, isolated by methods described here, is in agreement with published data (26).

DISCUSSION

These studies have shown that the bulk of DNA sequences complementary to both 18 S and 28 S ribosomal RNA were contained in the DNA of isolated nucleoli of liver cells. They therefore resolve the conflicting ideas about an extranucleolar synthesis of 18 S rRNA (12, 17, 20) and support the concept that the nucleolus is the site of synthesis of the two rRNAs (7). Although evidence has been obtained for the localization of DNA complementary to the two rRNAs in the nucleolar organizer region of the genome of *Drosophila melanogaster* (1, 2) and *Xenopus laevis* (3, 4), previous attempts to demonstrate this type of localization in isolated nucleoli have been unsatisfactory because of the relatively poor recovery of rRNA cistrons (5, 6).

The saturation plateaus for rRNAs of rat liver, as well as a number of other species previously examined (1–6, 26, 36), indicate that there is a multiplicity of rRNA cistrons in the

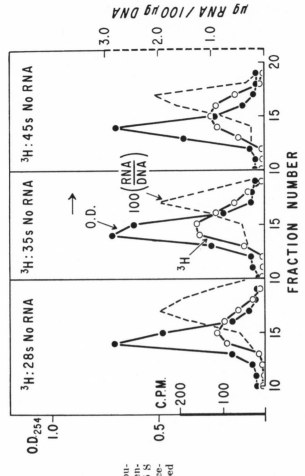

Fig. 7. Buoyant density distribution of nucleolar DNA complementary to ^{3}H-nucleolar 28 S RNA, 35 S RNA, and 45 S RNA. The procedures were the same as those described in the legend of Fig. 4.

genome. It may be calculated from the following information that there are about 300 sequences complementary to each of the two rRNAs per nucleolus: the fraction of nuclear DNA complementary to the two rRNAs (Fig. 1), the DNA content per nucleus (12 pg) and per nucleolus (0.37 pg), the average number of nucleoli per nucleus (2.4), a molecular weight of 0.63×10^6 and 1.6×10^6 for 18 S rRNA and 28 S rRNA, respectively, (35), and the assumption that one strand of DNA is copied for RNA synthesis.

Evidence for a contiguous array of rRNA cistrons in the genome has gained support from studies on the buoyant density of rRNA-DNA hybrids (3, 4, 37) and the distribution of rRNA cistrons in the DNA of bacteria (38) and *X. laevis* (3, 4), and now in a heavy satellite of the nucleolar DNA fraction of rat liver. This satellite accounts for about 0.6% of the total nuclear DNA and about 8% of the total nucleolar DNA. Since nucleolar DNA is equal to about 5.3×10^{11} daltons per nucleus and the combined fraction of nuclear DNA complementary to rRNA is equal to about 3.3×10^9 daltons, it may be calculated that the total rRNA cistrons account for only 0.6% of the total nucleolar DNA. This is consistent with the data obtained for the fraction of nucleolar DNA complementary to the two rRNAs (Fig. 2) that have been corrected for the noncoding strand of DNA. If it is assumed that all of the rRNA cistrons were contained in the heavy satellite, it may be calculated that they would comprise only 8% of the total heavy DNA satellite. This may account for the relatively low buoyant density (1.708 g per cm³) of the nucleolar heavy satellite compared to the high buoyant density (1.723 g per cm³) of the guanine- and cytosine-rich fragment of toad DNA (3), which is thought to contain large polycistronic clusters complementary to rRNA (4). It is possible that the relatively broad distribution found for the rRNA cistrons of the nucleolar DNA fraction (Fig. 3) is a consequence of random degradation of the DNA during isolation and purification.

In contrast to the early hypothesis that nucleolar 45 S RNA is cleaved to produce an 18 S rRNA and a 28 S rRNA (8, 9, 12), the data derived from molecular hybridization (Figs. 2 and 6) indicate that nucleolar 45 S RNA contains, in addition to the base sequences for the two rRNAs, a nonribosomal segment equal to about 35% of the total molecule (0.52 − (0.10 + 0.24)/ 0.52). The competition studies (Table II) showed that nucleolar 45 S rRNA shared a maximum of 80% of its base sequences in common with 18 S rRNA and 28 S rRNA. In view of the relatively high level of competition between 18 S rRNA and nucleolar 45 S RNA, 35 S RNA, and 28 S RNA, it is suggested that the nonribosomal segment of nucleolar 45 S RNA may share some of its base sequences in common with 18 S rRNA and that

184

this segment or a part of it may also be contained in both nucleolar 35 S RNA and 28 S RNA.[1] These studies also have some bearing on the relative size (molecular weight) of the preribosomal RNAs of the nucleolus. The saturation plateaus for nucleolar RNAs (Fig. 6) suggest that nucleolar 28 S RNA and 35 S RNA are similar in size, but smaller than nucleolar 45 S RNA and 55 S RNA by factors of 0.85 and 0.83, respectively. Whereas the transformation of nucleolar 55 S RNA to 45 S RNA may involve only a minor modification of primary structure, the 15% decrease in the saturation plateau (molecular size) between nucleolar 45 S RNA and 35 S RNA may be related to the proposed conversion of nucleolar 45 S RNA into an 18 S RNA and a 35 S RNA (19). Finally, it should be pointed out that nucleolar 28 S RNA appears to be markedly different from 28 S rRNA in size and over-all base sequence. Although these results are strongly suggestive of a precursor relationship among nucleolar 45 S RNA, 35 S RNA, and 28 S RNA and the two rRNAs, they do not clearly distinguish which one of the nucleolar RNAs is cleaved to produce the two rRNAs or their precursors.

The hypothesis that the transition of preribosomal RNAs to rRNAs involves a stepwise process of chain shortening (19) rather than changes in RNA tertiary structure (41) is also supported by the data on the base composition of nucleolar 45 S, 35 S, and 28 S RNAs (Table III). The progressive changes in the content of adenine and uracil of preribosomal RNAs suggest that pyrimidine nucleotides, especially uridylic acid, may be lost preferentially in the conversion of preribosomal nucleolar RNA to the two rRNAs. In addition, it is clear from a comparison of the data for the adenine content of the nucleolar RNAs and the two rRNAs that a conservative conversion of the preribosomal RNAs to the two rRNAs is not possible.

The disparity between the content of adenine and thymine of the nucleolar DNA fraction is interesting, since it suggests that the nucleolus may contain a small amount of single-stranded DNA. Such a DNA would be expected to form a zone at a

---

[1] Workers in a number of other laboratories (6, 8, 39, 40) have investigated the competition between rRNAs and rapidly sedimenting RNA fractions for sites on the DNA of whole cells and isolated nucleoli. The interpretation of the data is difficult, however, because of the heterogeneity with respect to sedimentation coefficient of the rapidly sedimenting RNAs used, the presence of contaminating rRNAs in RNA fractions isolated from whole cells and nuclei (12), and finally the spurious competition found here and previously (36) between the two rRNAs when present together at saturating amounts.

higher buoyant density in CsCl than native double-stranded DNA (42), and thus it may comprise a part of the heavy DNA satellite of the nucleolar fraction described here. Clearly, the small contribution (0.6%) of the guanine- and cytosine-rich rRNA cistrons of the nucleolar DNA fraction could not account for the magnitude of this disparity.

Acknowledgments—I wish to thank Drs. R. Pruitt and J. Schofield for financial support to complete this work, Dr. M. Mandel for analytical ultracentrifuge analyses, and Dr. S. Black for help with *B. megaterium* cultures. I also wish to thank Miss K. Byron and Mr. J. L. Hodnett for excellent technical assistance in the early phases of this work.

REFERENCES

1. RITOSSA, F. M., AND SPIEGELMAN, S., *Proc. Nat. Acad. Sci. U. S. A.*, **53**, 737 (1965).
2. RITOSSA, F., ATWOOD, K., LINDSLEY, D., AND SPIEGELMAN, S., *Nat. Cancer Inst. Monogr.*, **23**, 449 (1966).
3. WALLACE, H., AND BIRNSTIEL, M. L., *Biochim. Biophys. Acta*, **114**, 296 (1966).
4. BIRNSTIEL, M. L., WALLACE, H., SIRLIN, J. L., AND FISCHBERG, M., *Nat. Cancer Inst. Monogr.*, **23**, 431 (1966).
5. CHIPCHASE, M. I. H., AND BIRNSTIEL, M. L., *Proc. Nat. Acad. Sci. U. S. A.*, **50**, 1101 (1963).
6. McCONKEY, E. H., AND HOPKINS, J. W., *Proc. Nat. Acad. Sci. U. S. A.*, **51**, 1197 (1964).
7. PERRY, R. P., *Proc. Nat. Acad. Sci. U. S. A.*, **48**, 2179 (1962).
8. SCHERRER, K., LATHAM, H., AND DARNELL, J. E., *Proc. Nat. Acad. Sci. U. S. A.*, **49**, 240 (1963).
9. PERRY, R. P., *Nat. Cancer Inst. Monogr.*, **14**, 73 (1964).
10. STEELE, W. J., OKAMURA, N., AND BUSCH, H., *J. Biol. Chem.*, **240**, 1742 (1965).
11. MURAMATSU, M., AND BUSCH, H., *J. Biol. Chem.*, **240**, 3960 (1965).
12. MURAMATSU, M., HODNETT, J. L., STEELE, W. J., AND BUSCH, H., *Biochim. Biophys. Acta*, **123**, 116 (1966).
13. PENMAN, S., *J. Mol. Biol.*, **17**, 117 (1966).
14. GREENBERG, H., AND PENMAN, S., *J. Mol. Biol.*, **21**, 527 (1966).
15. ZIMMERMAN, E. F., AND HOLLER, B. W., *J. Mol. Biol.*, **23**, 149 (1967).
16. GIRARD, M., LATHAM, H., PENMAN, S., AND DARNELL, J. E., *J. Mol. Biol.*, **11**, 187 (1965).
17. STEELE, W. J., AND BUSCH, H., *Biochim. Biophys. Acta*, **129**, 54 (1966).
18. PENMAN, S., SMITH, I., HOLTZMAN, E., AND GREENBERG, H., *Nat. Cancer Inst. Monogr.*, **23**, 489 (1966).
19. WEINBERG, R. A., LOENING, U., WILLEMS, M., AND PENMAN, S., *Proc. Nat. Acad. Sci. U. S. A.*, **58**, 1088 (1967).
20. STEELE, W. J., *Proceedings of the International Symposium on the Cell Nucleus, Rijswijk, The Netherlands, 1966*, Taylor and Francis, Ltd., London, 1966, p. 203.
21. MURAMATSU, M., SMETANA, K., AND BUSCH, H., *Cancer Res.*, **23**, 510 (1963).

22. STEELE, W. J., AND BUSCH, H., *Biochim. Biophys. Acta*, **119**, 501 (1966).
23. CHAUVEAU, J., MOULÉ, Y., AND ROUILLER, C., *Exp. Cell Res.*, **11**, 317 (1956).
24. KIRBY, K. S., *Biochem. J.*, **66**, 495 (1957).
25. KIRBY, K. S., *Biochem. J.*, **96**, 266 (1965).
26. YANKOFSKY, S. A., AND SPIEGELMAN, S., *Proc. Nat. Acad. Sci. U. S. A.*, **49**, 538 (1963).
27. GILLESPIE, D., AND SPIEGELMAN, S., *J. Mol. Biol.*, **12**, 829 (1965).
28. MESELSON, M., STAHL, F. W., AND VINOGRAD, J., *Proc. Nat. Acad. Sci. U. S. A.*, **43**, 581 (1957).
29. WETTSTEIN, F. O., STAEHELIN, T., AND NOLL, H., *Nature*, **197**, 430 (1963).
30. HENSHAW, E. C., *J. Mol. Biol.*, **9**, 610 (1964).
31. STEELE, W. J., AND BUSCH, H., in H. BUSCH (Editor), *Methods in cancer research, Vol. III*, Academic Press, New York, 1967, p. 61.
32. NYGAARD, A. P., AND HALL, B. D., *Biochem. Biophys. Res. Commun.*, **12**, 98 (1963).
33. MARSHAK, A., AND VOGEL, H. J., *J. Biol. Chem.*, **189**, 597 (1951).
34. BEAVEN, G. H., HOLIDAY, E. R., AND JOHNSON, E. A., in E. CHARGAFF AND J. N. DAVIDSON (Editors), *The nucleic acids, Vol. I*, Academic Press, New York, 1955, p. 498.
35. PETERMANN, M. L., AND PAVLOVEC, A., *Biochim. Biophys. Acta*, **114**, 264 (1966).
36. ATTARDI, G., HUANG, P., AND KABAT, S., *Proc. Nat. Acad. Sci. U. S. A.*, **54**, 185 (1965).
37. YANKOFSKY, S. A., AND SPIEGELMAN, S., *Proc. Nat. Acad. Sci. U. S. A.*, **48**, 1466 (1962).
38. DAVISON, P. F., *Science*, **152**, 509 (1966).
39. PERRY, R. P., SRINIVASAN, P. R., AND KELLEY, D. E., *Science*, **145**, 504 (1964).
40. YOSHIKAWA-FUKADA, M., *Biochim. Biophys. Acta*, **123**, 91 (1966).
41. MURAMATSU, M., HODNETT, J. L., AND BUSCH, H., *J. Biol. Chem.*, **241**, 1544 (1966).
42. DOTY, P., MARMUR, J., EIGNER, J., AND SCHILDKRAUT, C., *Proc. Nat. Acad. Sci. U. S. A.*, **46**, 461 (1960).

Ribosomes in Rat Liver:
An Estimate of the Percentage of Free and Membrane-bound Ribosomes interacting with Messenger RNA *in vivo*

Ribosomes interact with mRNA to form polysomes (Warner, Rich & Hall, 1962; Wettstein, Staehelin & Noll, 1963; Gierer, 1963). How many ribosomes in rat liver are engaged with mRNA in polysomal aggregates is not known. The conventionally used procedure of Wettstein *et al.* (1963) for the isolation of polysomes sediments and analyzes only two-thirds of the ribosomes present in the postmitochondrial supernatant fraction (Blobel & Potter, 1967). Whether the non-sedimented ribosomes also occur as polysomes is not known. Complete recovery by longer centrifugation, on the other hand, increases the chances for nucleolytic attack and breakdown of polysomes.

It was shown that an RNase inhibitor occurring in the high-speed supernatant fraction of rat liver (Roth, 1956) completely inhibits the breakdown of polysomes in the presence of endogenous and exogenous RNase (Blobel & Potter, 1966). The use of this RNase inhibitor throughout the cell fractionation procedure permitted, in addition to complete recovery, a separation of free and membrane-bound ribosomes intact as polysomes. From the sedimentation profiles of the separated free and membrane-bound ribosomes in sucrose gradients, an estimate has been made of the fraction of ribosomes (a) interacting with mRNA to form polysomes, (b) occurring as monomers(somes) and dimers(somes), and (c) dissociated into subunits.

150- to 200-g Holtzman rats (Holtzman Co., Madison, Wisconsin) maintained on a Purina chow diet were fasted from 9.00 p.m. to 9.00 a.m. and then decapitated with a guillotine (Harvard Apparatus Co., Dover, Mass.). The livers were removed and chilled in ice-cold 0·25 M-sucrose in TKM (0·05 M-Tris–HCl, pH 7·5 at 20°C, 0·025 M-KCl, 0·005 M-MgCl$_2$). All subsequent operations were performed at temperatures near 0°C. After being minced with scissors, the livers were homogenized with 10 strokes in two volumes 0·25 M-sucrose in TKM in a Potter–Elvehjem homogenizer with Teflon pestle (clearance: 0·010 in.). A postmitochondrial supernatant (S$_2$) was prepared by centrifuging the homogenate in the SS 34 rotor in the Servall centrifuge for 10 minutes at 17,000 g (max.). The postmitochondrial supernatant fraction was centrifuged in a Spinco 40 rotor for four hours at 40,000 rev./min (105,000 g average), and the resulting high-speed supernatant fraction (S$_3$) was frozen at − 20°C. Storage in the frozen state retained the RNase inhibitor activity unchanged for at least a month (Roth, 1958).

A postmitochondrial supernatant fraction (S$_2$) was obtained as described above, except that the liver was homogenized in a sucrose–S$_3$ medium (9 parts 0·25 M-sucrose in TKM and 1 part S$_3$). 4 ml. S$_2$ was layered over a two-layer discontinuous sucrose gradient with 3 ml. 2·0 M-sucrose–S$_3$ medium at the bottom and 3 ml. 1·38 M-sucrose–S$_3$ medium at the top of a Polyallomer tube. Both sucrose media were prepared from a stock solution of 2·3 M-sucrose in TKM, which had been filtered through a Millipore filter (1·2 μ) and was kept frozen at − 20°C. To 10 ml.

of this stock solution 1·5 ml. S_3 was added to give 2·0 M-sucrose–S_3 medium. The 1·38 M-sucrose–S_3 medium was prepared by adding 8 ml. S_3 to 10 ml. stock solution. The gradients were centrifuged for 24 hours at 40,000 rev./min in a Spinco 40 rotor. Under these conditions, all free ribosomes in the S_2 sedimented into a pellet (Blobel & Potter, 1967). The pellets were stored at 0°C for 24 hours, to be analyzed at the same time as the bound ribosomes for their sedimentation profiles in sucrose gradients. (There was no detectable difference between these profiles and those obtained without storage of the pellets.) The ribosome–membrane complex (rough endoplasmic reticulum) sedimented into the 1·38 M-sucrose medium in the form of many discrete bands. (If the 0·5 M-sucrose of the original procedure of Wettstein et al. (1963) were used, the rough endoplasmic reticulum would accumulate in a small band on top of the 2·0 M-sucrose, with some precipitation at the wall of the tube, so that quantitative recovery of this fraction would be difficult.) The 1·38 M-layer containing the rough endoplasmic reticulum was removed with a syringe (after the 4 ml. above this layer had been taken off with a syringe and discarded). After rehomogenization, a detergent mixture of Triton X-100 (20% w/v) and sodium deoxycholate (5% w/v) was added to give a final concentration of 4% Triton X-100 and 1% sodium deoxycholate. The mixture was underlaid with 3 ml. of 2·0 M-sucrose–S_3 medium and again spun for 24 hours at 40,000 rev./min. in a Spinco 40 rotor. This pellet contained the ribosomes which were originally bound to the membranes and which are, therefore, referred to as membrane-bound ribosomes. Both free and membrane-bound ribosomes were obtained together when, under the above conditions, the detergent mixture was added to the postmitochondrial supernatant fraction.

The sedimentation profiles of the various ribosome preparations are shown in Fig. 1. The quantitative analysis of the area under the profile is recorded in Table 1. The area to the left of peak 3 (between the dotted lines in Fig. 1) represents polysomes or ribosomes interacting with mRNA. However, some of the heavy polysomes sediment to the bottom of the tube. A correction for this loss was added to the polysome segment by extrapolation of the profile to the base line. It is not possible to correct for the sedimentation of heavy polysomes by simply adding the value for all the sedimented ribosomes to the value for the polysome region. It was found that the sedimented ribosomes, which amount to as much as 30% of the layered ribosomes, after resuspension in the presence of RNase inhibitor, show a sedimentation profile similar to their parental profiles. The sedimented ribosomes are therefore representative of all segments of the parental profile which, on hitting the wall of the non-sectorial centrifuge tube, move down into the pellet ("wall effect", Anderson, 1955). The middle segment of the sedimentation profile (dashed lines) represents single or double ribosomes which might or might not interact with mRNA. Since it is known that monomers can dimerize in high magnesium concentration (Petermann, 1964), it could be argued that a similar dimerization might occur between monomers and ribosomes engaged with mRNA in the polysome region to form pseudo-polysomes. This would lead to a serious underestimation of the amount of monomers in favor of polysomes. However, we found that the sedimentation profiles in 0·001 M-magnesium (instead of 0·005 M-magnesium) are identical, except for a slightly smaller dimer peak and a correspondingly higher monomer peak. The small segment to the right delineates the area under the two ribosomal subunit peaks.

From Table 1 it can be seen that 85% of the free ribosomes (Fig. 1(a)) were interacting with mRNA to form polysomes. It has not been determined how many of the

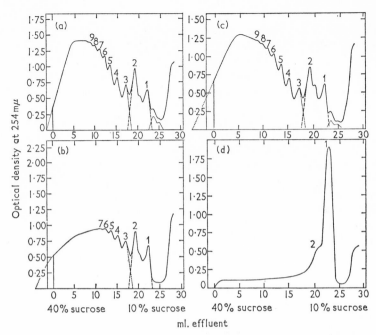

Fig. 1. Sedimentation profiles of rat liver ribosomes in sucrose gradients.

Ribosomal pellets were resuspended in 2 ml. water and 0·2 ml. S_3 by frequent stirring on a Vortex mixer with a spatula in the tube. 0·6 ml. of the ribosome–S_3 solution containing approximately 40 optical density units was layered on 28 ml. of a linear sucrose gradient (10 to 40% w/v in TKM) and spun for 2 hr (without subsequent use of the brake) in a Spinco SW25 rotor at 25,000 rev./min. The optical density was monitored as described recently (Blobel & Potter, 1966).

(a) Free ribosomes; (b) bound ribosomes; (c) free and bound ribosomes. 0·15 ml. of each of the three ribosome–S_3 solutions was treated with 2 μg RNase type XIIA (Sigma Chemical Co.) at 0°C and immediately thereafter layered. The three sedimentation profiles obtained were similar, and (d) represents the profile of RNase-treated free ribosomes.

TABLE 1

Distribution of polysomes, monoribosomes, dimers and subunits in rat liver

Area	Free ribosomes (100%)	Bound ribosomes (100%)	Free and bound ribosomes (100%)
	Per cent	Per cent	Per cent
Polysomes	85	85	87
Peaks 1 and 2	13	15	12
Subunits	2	0	1

Based on areas under the sedimentation profiles of Fig. 1(a), (b) and (c). Numbers are percentages of the total area comprising polysomes (between dotted lines), peaks 1 and 2 (between dashed lines) and subunits (under solid lines).

190

ribosomes in the middle segment were monosomes and disomes in combination with mRNA, but evidence from other sources indicates, but does not prove, that mRNA may be absent from both of these fractions *in vivo* (Webb & Potter, 1966). Discernible subunit peaks were found only in the profiles of free ribosomes and amounted to approximately 2%. It was shown recently (Blobel & Potter, 1967) that the post-mitochondrial supernatant fraction contains more than 90% of the free ribosomes but only about 30% of the bound ribosomes present in the homogenate. Since Lawford, Langford & Schachter (1966) prepared polysomes also from the pellet fractions usually discarded when the postmitochondrial supernatant is prepared, the above results with the bound ribosomes of the postmitochondrial supernatant fraction can be extrapolated safely to the total membrane-bound ribosomes. Inasmuch as 25% of the total ribosome population is free and 75% is membrane-bound (Blobel & Potter, 1967), we conclude that at least 85% of the cellular ribosomes in rat liver are in combination with mRNA, regardless of whether they are free or membrane-bound. The ribosomal subunits amount to less than 1% of the total ribosome population and do not occur membrane-bound. Approximately 15% of the total ribosomes occur as monomers (somes) or dimers (somes) with the extent of the interaction with mRNA remaining unknown.

Quantitative data of this type have been reported for bacteria (Mangiarotti & Schlessinger, 1966). In sharp contrast to our findings with rat liver, they reported a large fraction of ribosomes dissociated into subunits, with the rest organized into polysomes.

One of us (G. B.) is a U.S. Public Health Service Postdoctoral Fellow. This work was supported by departmental grant CA-07175 from the National Cancer Institute, U.S. Public Health Service.

Günter Blobel
Van R. Potter

REFERENCES

Anderson, N. G. (1955). *Exp. Cell Research*, **9**, 446.
Blobel, G. & Potter, V. R. (1966). *Proc. Nat. Acad. Sci., Wash.* **55**, 1238.
Blobel, G. & Potter, V. R. (1967). *J. Mol. Biol.* **26**, 279.
Gierer, A. (1963). *J. Mol. Biol.* **6**, 148.
Lawford, G. R., Langford, P. & Schachter, H. (1966). *J. Biol. Chem.* **241**, 1835.
Mangiarotti, G. & Schlessinger, D. (1966). *J. Mol. Biol.* **20**, 123.
Petermann, M. (1964). In *The Physical and Chemical Properties of Ribosomes*. Amsterdam: Elsevier Publishing Co.
Roth, J. S. (1956). *Biochim. biophys. Acta*, **21**, 34.
Roth, J. S. (1958). *J. Biol. Chem.* **231**, 1085.
Warner, J. R., Rich, A. & Hall, C. E. (1962). *Science*, **138**, 1399.
Webb, T. E. & Potter, V. R. (1966). *Cancer Research*, **26**, 1022.
Wettstein, F., Staehelin, T. & Noll, H. (1963). *Nature*, **197**, 430.

ACRYLAMIDE GEL ELECTROPHORESIS OF
HELA CELL NUCLEOLAR RNA*

By Robert A. Weinberg, Ulrich Loening, Margherita Willems,
and Sheldon Penman

The nucleolus of eucaryotic cells has been identified by Perry and others[1-4] as the site of ribosomal RNA synthesis. It has recently become possible to obtain from HeLa cells a nucleolar preparation which, as seen by electron microscopy, is relatively free of chromatin.[5, 6] Fractionation of C^{14}-uridine-labeled cells confirmed the hypothesis that the nucleolus is the site of synthesis of the 45S ribosomal RNA precursor.[5] Another species of RNA (32S) is present in relatively large amounts. The nucleolus appears to contain only ribosomal precursor RNA since if fractionation is performed carefully, very little of the nucleoplasmic heterodisperse RNA is associated with it.[7, 8] The following picture of the major events in ribosomal RNA formation has emerged. The initial event is the synthesis of a high-molecular-weight precursor molecule with a sedimentation constant of 45S.[9-11] After 15 to 20 minutes, this molecule is cleaved, yielding 18S ribosomal RNA and a species of RNA whose sedimentation constant is 32S.[12] The 18S RNA is quickly transported from the nucleolus and appears in the cytoplasm as part of the smaller ribosomal subunit. After additional processing time in the nucleolus, the 32S molecule is converted to 28S and eventually emerges into the cytoplasm as part of the larger ribosomal subunit.

With the development by Loening of a method of polyacrylamide gel electrophoresis for molecules as large as ribosomal RNA,[13] it has become possible to study in greater detail the events of nucleolar RNA processing. The gels used in these experiments have been modified by the addition of glycerol to facilitate freezing and slicing.

The following information has been obtained concerning nucleolar processing of ribosomal RNA. (1) The site of transformation of 32S to 28S RNA is the nucleolus. (2) Several additional short-lived intermediate species of ribosomal RNA have been identified with estimated sedimentation constants of 41S, 36S, and 20S. (3) Some short-lived intermediates increase in amount under conditions that disrupt normal nucleolar RNA processing, e.g., poliovirus infection. (4) The conversion of 45S RNA to mature ribosomal RNA is accompanied by a net loss of RNA. This is also shown for the transformation of 32S RNA to 28S.

Materials and Methods.—Cells: HeLa type 3 cells were grown and labeled in suspension culture as previously described.[14]

Radioisotopes: L-methionine-methyl-C^{14} (49 mc/mM) and uridine-2-C^{14} (27 mc/mM) were purchased from Schwarz BioResearch. L-methionine-methyl-H^3 (1400 mc/mM) was purchased from Nuclear Chicago. Methyl labeling was performed in Eagle's medium free of unlabeled methionine and containing adenosine and guanosine (2×10^{-5} M). When noted, unlabeled methionine was added back to this medium to prevent methionine starvation.

Cell fractionation: Cells were separated into nuclear and cytoplasmic fractions as previously described.[11] The cleaned nuclei from approximately 4×10^7 cells were suspended in 1 ml of high-ionic-strength buffer (HSB: 0.5 M NaCl, 0.05 M MgCl₂, 0.01 M Tris, pH 7.4), warmed briefly to 37°, and digested with 50 μg of Worthington electrophoretically purified DNase. The digest

192

was layered on a 16-ml sucrose gradient [15–30% (w/w) sucrose in HSB] and centrifuged 15 min at 22,000 rpm at 5°C in the SW 25.3 rotor of the Spinco model L-2 ultracentrifuge. The pellet was termed nucleoli and the supernatant fraction termed nucleoplasm.

RNA extraction: RNA was extracted with hot phenol-SDS (sodium dodecyl sulfate) as described before.[9, 11] The extracted RNA was precipitated with 2 to 3 volumes of 95% ethanol after addition of yeast tRNA (2 optical density units/ml) as carrier (Mann Research Co.). Precipitates were collected by centrifugation for 15 min at 15,000 rpm in the International model B-20. Pellets were suspended in 50 μl of electrophoresis buffer (0.04 M Tris, 0.02 M sodium acetate, 2 mM EDTA, 0.5% SDS, adjusted to pH 7.4 with acetic acid) made 15% (v/v) in glycerol. This solution was layered directly onto the polyacrylamide gel.

Polyacrylamide gels: Gels were polymerized[13] in a buffer identical to the above electrophoresis buffer except that the SDS was absent and the glycerol concentration was 10% (v/v). The polymerizing mixture was made 2.7% (w/v) in acrylamide (purchased from Eastman Chemical Co. and recrystallized from chloroform) and 0.25% (v/v) in the alkali-labile cross-linking agent ethylene diacrylate[15] (K & K laboratories). Acrylamide solutions and glycerol were shaken with hexane at room temperature to remove residual material absorbing at 2600 Å. Cylindrical gels of 0.6 cm diameter and 6 cm length were cast. Electrophoresis tanks were filled with electrophoresis buffer containing the SDS and 10% (v/v) glycerol. Electrophoresis was performed at 5 ma per gel for the period specified. The gels were scanned for absorbancy with a Gilford recording spectrophotometer adapted for this purpose. Gels containing radioactivity were subsequently frozen and sliced in 1-mm slices in a bath of hexane adjusted to about −30°C by the addition of dry ice. The glycerol contained in the gel prevents ice crystal formation upon freezing and therefore facilitates slicing. Gel slices to be counted by scintillation were hydrolyzed in vials with 0.5 ml of concentrated NH$_4$OH for 1 hr. Ten ml of Bray's[16] scintillation fluid were added. Slices to be counted in a gas-flow system were placed on planchettes adapted for this purpose and hydrolyzed with 1 ml of concentrated NH$_4$OH, which was subsequently evaporated by heating.

Results.—The greatly improved resolution possible with acrylamide gel electrophoresis compared to sucrose density gradients is demonstrated in Figure 1. RNA

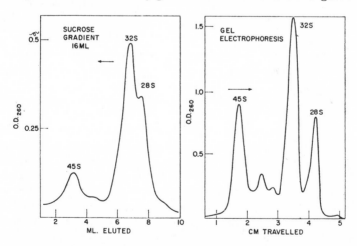

FIG. 1.—Comparison of sucrose gradient and gel electrophoresis analysis of nucleolar RNA. Each frame represents nucleolar RNA from approximately 4 × 10[7] cells. *Sucrose gradient:* Sample resuspended in 0.4 ml of SDS buffer and run for 14 hr at 23,000 rpm. The gradient was 15–30% (w/w) sucrose in SDS buffer (0.1 M NaCl, 0.01 M Tris pH 7.4, 0.001 M EDTA, 0.5% sodium dodecyl sulfate). The total volume was 16 ml. Centrifugation was in the SW 25.3 rotor of the L-2 Spinco ultracentrifuge. The bottom two-thirds of the elution pattern is displayed here. *Gel electrophoresis:* The sample was run for 6 hr. The top two-thirds of gel is displayed here.

from HeLa cell nucleolar preparations was analyzed on both a sucrose gradient and an acrylamide gel. The optical density profiles of both are shown. The principal components of nucleolar RNA are visible in the pattern obtained from the sucrose gradient. These are 45S and 32S RNA as well as a shoulder corresponding to a 28S RNA component and a suggestion of ultraviolet-absorbing material sedimenting between 45S and 32S RNA. These additional species of.RNA are clearly resolved by gel electrophoresis. In addition to 45S and 32S RNA, there is a well-separated peak corresponding to 28S ribosomal RNA. There is an additional peak visible which corresponds to RNA which would sediment at approximately 41S as well as a very minor component at approximately 36S. The identification of the RNA species observed in the gels was confirmed by electrophoretic analysis of the RNA species isolated from conventional 15–30 per cent sucrose gradient zones.

When electrophoresis is carried out for a shorter period, two additional very small peaks of absorbing material are visible which correspond to 18S and 20S. The profile of nucleolar RNA obtained after four hours of electrophoretic migration is shown in Figure 2. The nucleolar RNA had been labeled with methionine-me-

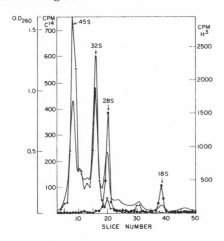

FIG. 2.—Electrophoretic analysis of nucleolar and ribosomal RNA. One hundred ml of cells at a concentration of 4×10^5 cells/ml were centrifuged and resuspended for 30 min in 20 ml of methionine-free medium containing 10 μc of C^{14}-methione. Nucleolar RNA was run for 4 hr with H^3-uridine-labeled cytoplasmic ribosomal RNA as marker. O.D., ———; C^{14}-methionine, ●-●-●; H^3-uridine, ■-■-■.

thyl-C^{14} for 30 minutes prior to fractionation. It has been shown previously that the only high-molecular-weight RNA that is methylated in short pulses is the 45S precursor and that methylation takes place shortly after synthesis.[17, 18] By 30 minutes, some of the labeled 45S has been processed and radioactivity has begun to enter the 32S RNA and 20S peaks. Tritium-labeled ribosomal RNA of high specific activity was added as a marker and served to identify the 18S and 28S nucleolar RNA.

After 30 minutes, all species of nucleolar RNA are labeled by methionine, with the exception of 28S and 18S. This finding is consistent with the view that the minor components are intermediates in the formation of mature ribosomal RNA. It is tentatively assumed that the 41S and 36S RNA are precursors of 32S and that 20S RNA is the precursor of 18S.

The very small amounts of 41S, 36S, and 20S RNA found in the nucleolus indicate that their lifetime is in the order of two minutes. This short lifetime, coupled with possible dispersion in the processing time of individual RNA molecules, prevents measurement of precise kinetics of formation of these minor components. The

194

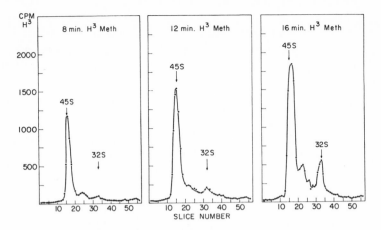

FIG. 3.—Kinetics of labeling of nucleolar RNA. Three hundred ml of cells were centrifuged and resuspended in 30 ml of methionine-free medium. Cells were labeled for 8, 12, and 16 min with 10 μc/ml of H³-methionine. Nucleolar RNA was run for 5 hr.

following experiment indicates that 41S and 36S are labeled after the 45S RNA and, in the presence of actinomycin D, decay after the disappearance of 45S. The results of short periods of incorporation of methyl-labeled methionine are shown in Figure 3. It can be seen that even after 12 minutes of incorporation, only the 45S species has appreciable radioactivity. By 16 minutes, however, the 41S and 36S RNA are apparently nearly fully labeled and radioactivity has begun to appear in the 32S region.

The decay of the 41S and 36S species in actinomycin is shown in Figure 4. Cells were labeled for 16 minutes with methionine and then actinomycin was added to the culture. Nine minutes after the addition of actinomycin, the 45S RNA radioactivity has decayed to approximately one third of its initial value. However, the radioactivity remains associated with 41S-36S RNA. By 26 minutes, all the radioactivity in 45S, 41S, and 36S peaks has disappeared and has apparently largely migrated to the 32S peak in the nucleolus and 18S in the cytoplasm.[11] The pattern of radioactivity after 26 minutes in actinomycin also shows the first appearance of radioactivity in the nucleolar 28S species of RNA.

The minor components of nucleolar RNA are difficult to analyze because of their

FIG. 4.—Effect of actinomycin on prelabeled nucleolar RNA. Three hundred ml of cells were centrifuged and resuspended in 25 ml of methionine-free medium. Cells were labeled for 16 min with 12 μc of C¹⁴-methionine and a sample was taken. Actinomycin D was then added to a concentration of 10 μg/ml. Samples were taken at 9 and 26 min after the addition of actinomycin. Nucleolar RNA was run for 7 hr.

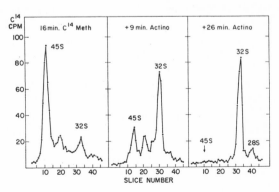

extremely low concentration. It is possible, however, to affect cell metabolism so that the content of the various species of RNA is altered and the concentration of some of the minor species is increased. One means of altering nucleolar RNA content is by infection with poliovirus. HeLa cells are unusual in that infection with small RNA viruses such as poliovirus does not result in the abrupt cessation of host RNA synthesis.[19] Rather, host-cell-directed RNA synthesis continues until quite late in infection but the processing of RNA, especially in the nucleolus, becomes aberrant shortly after infection. It was first noted by Willems that various intermediates apparently accumulate in the nucleolus.[20] In the experiment whose results are shown in Figure 5, HeLa cells were infected with poliovirus at a multiplicity of 500

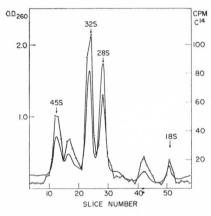

Fig. 5.—Effect of poliovirus infection on nucleolar contents. One hundred ml of cells at 4×10^5 cells/ml were centrifuged and resuspended in 3 ml of serum-free Eagle's medium containing 5×10^9 PFU/ml of poliovirus and 2 mM guanidine. After $1/2$ hr absorption, cells were centrifuged and resuspended in 20 ml of medium containing 3 μc of C^{14}-methyl methionine, 5×10^{-5} M cold methionine, and 2 mM guanidine. Cells were fractionated after 90 min of labeling. Nucleolar RNA was run for 5 hr. O.D., ——; C^{14}-methionine, –O–O.

PFU per cell. Thirty minutes after infection, C^{14}-methionine was added and incorporation allowed to continue for 90 minutes. The nucleolar content of the minor RNA species, 41S, 28S, 20S, and 18S, is seen to be significantly increased when compared to the profile seen in Figure 2. This increased content is apparently due to aberrant processing induced by poliovirus infection. It should be noted that these experiments were carried out in the presence of 2×10^{-3} M guanidine, which prevents viral RNA replication[21] but permits the expression of several virus-directed effects on host cell metabolism including the suppression of host cell protein synthesis[22] and the disarrangement of nucleolar processing shown here.[7] A detailed report on further experiments of this type will appear elsewhere.

The high resolution afforded by the acrylamide gel electrophoresis technique makes another type of measurement possible. Cells can be labeled with both methionine and uridine so that all species of RNA in the nucleolus have been uniformly labeled. It is then possible to measure the uridine and methyl group content of the various species of nucleolar RNA. This type of experiment could not be done previously because of the extensive cross-contamination of the various species of RNA obtained in sucrose gradients. It has been shown that the 45S ribosomal precursor RNA is methylated while it is being polymerized on the DNA template,[17] and that virtually all methylation takes place on the 45S molecule.[17, 18] Furthermore, the detailed pattern of methylation of the 45S precursor RNA is shown to be identical to that of 18S plus 28S RNA.[23] It appears, therefore, that virtually all methyl groups destined to be associated with 18S and 28S RNA are already attached to the 45S.

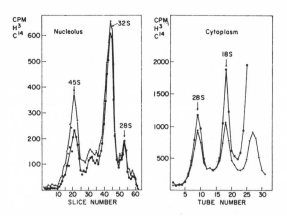

FIG. 6.—Comparison of uridine and methyl labeling of nucleolar and ribosomal RNA. One hundred ml of cells were centrifuged and resuspended in 50 ml of methionine-free medium to which were added 0.5 mg of cold methionine, 0.72 mg of cold uridine, 400 μc of H³-methionine, and 1.5 μc of C¹⁴-uridine. The cells were labeled for 2.5 hr and the nucleoli run for 7 hr. The cytoplasm was centrifuged for 15 hr at 24,000 rpm through a 28 ml 15–30% sucrose gradient in SDS buffer on the SW 25 rotor of the Spinco model L ultracentrifuge. C¹⁴-uridine, —●—●—; H³ - methionine, —■—■—.

Previous experimental results imply that there is little loss of methyl label as the 45S precursor is processed to form the mature 28S and 18S components of ribosomes.[23] The experimental results shown in Figure 6 show a uridine-to-methionine ratio which is highest for 45S and decreases for 32S, 28S, and 18S. This implies the loss of uridine- but not methionine-labeled RNA during processing. Cells were labeled for 2.5 hours with C¹⁴-uridine and H³-methionine. The specific activity and concentration of radioactive precursor was adjusted so that there was little change in the rate of uptake during the period of incorporation. RNA extracted from the nucleolar fraction was analyzed by gel electrophoresis. The total RNA from the cytoplasm was analyzed on a sucrose gradient. The label specifically associated with each peak was estimated and the results are shown in Table 1.

Discussion.—The electrophoretic separation of RNA on acrylamide gels makes possible a degree of resolution heretofore unobtainable and permits an extremely detailed analysis of the events in ribosomal RNA processing. Only part of the 45S precursor is converted to 28S and 18S end products. The fate of the remaining portion is unknown at this time. If the reduction in molecular weight proceeds stepwise, then it would be expected that short-lived intermediates would be observed. Several such intermediates have been identified in these experiments and their proposed role is shown in Figure 7.

It would be desirable to know whether the formation of 18S RNA or its possible 20S precursor accompanies the formation of 41S intermediate RNA, or whether they are products of the subsequent cleavage of this molecule. Because of the very short lifetime of the intermediates, it has not been possible to establish unambiguously whether 18S or 20S RNA appear coincident with or subsequent to the formation of 41S RNA.

TABLE 1

RATIO OF C¹⁴-URIDINE LABEL TO H³-METHYL
METHIONINE LABEL FOR RNA OF VARIOUS TYPES

Nucleolus		Cytoplasm		Weighted av. of
45S	32S	28S	18S	28S and 18S
1.59	1.16	0.79	0.55	0.75

The radioactivity derived from C¹⁴-uridine and H³-methyl methionine associated with each species of RNA in Figure 6 was estimated and the ratio calculated. The weighted average of 18S and 28S was calculated on the assumption that they have molecular weights of 6 × 10⁵ and 1.6 × 10⁶ daltons, respectively.[26]

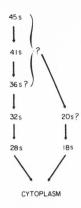

45s

41s   ?

36s?

32s   20s?

28s   18s

CYTOPLASM

Fig. 7.—Proposed metabolic relation of the various species of nucleolar RNA.

Two separate kinds of evidence support the conclusion that there is a loss of molecular weight in the processing of the 45S ribosomal precursor into its final products. The first of these concerns the conversion of 32S into 28S RNA. It is observed that the electrophoretic mobility of 28S RNA is higher than that of 32S RNA. However, the 32S RNA sediments faster than 28S RNA. If 28S RNA represented a conformational change of 32S RNA, then it should be a more expanded conformation to account for its slower sedimentation constant. If 28S were more expanded, then its electrophoretic mobility should be lower than 32S since the separation in this type of acrylamide gel should be primarily on the basis of size. Since the 28S, in fact, migrates in gels more rapidly than 32S, it cannot be a more expanded form of that molecule and hence must represent a structure of smaller molecular weight.

Other evidence for nonconservative conversions is derived from comparison of C^{14}-uridine/H^3-methyl ratios as shown in Figure 6 and Table 1. On the assumption that methyl label is conserved during processing, the observed change in the ratio of uridine to methyl label is consistent with the loss of approximately half of the 45S precursor molecule during processing. Assuming that uridine labeling is roughly proportional to molecular weight, and that the molecular weights of the 18S and 28S are approximately 0.6×10^6 and 1.6×10^6,[26] we estimate that the molecular weights of the 45S and 32S RNA are 4.6×10^6 and 2.3×10^6 daltons, respectively. The molecular weights, estimated by assuming that $S \propto M^{0.5}$,[24] are 4.1×10^6 daltons for the 45S and 2.1×10^6 daltons for 32S. These determinations are extremely crude and only indicate the nonconservative nature of the transitions.

The nonconservative nature of the processing of 45S has also been indicated in experiments in which the base composition of 45S RNA has been compared to that of its various products. A change in base composition is observed which is consistent with the loss of a significant fraction of the 45S precursor molecule.[27]

The experiment in which nucleolar content is altered by poliovirus infection indicates that the various minor components found by gel electrophoresis are not artifacts since it is unlikely that alterations in cell metabolism could affect the physicochemical processes responsible for artifacts during the extraction procedure.

While precise kinetics on the formation and decay of the various intermediates are not at present measurable, it has been demonstrated that the 41S and 36S RNA are formed after 45S and before 32S RNA. Similarly, other experiments have shown that the 20S peak is labeled after 45S and, as shown in Figure 2, before the 18S nucleolar components. On the assumption that all these RNA species are, in fact, involved in the processing of ribosomal RNA and do not represent aberrant forms, a scheme incorporating the observations reported here is shown in Figure 7. The question marks indicate that the kinetic data on 41S, 36S, and 20S RNA are not at present sufficient to establish the unambiguous role of these species and, as mentioned previously, the stage at which the 20S precursor is formed is not known. The scheme in Figure 7 should then be taken as summarizing our knowledge as it now exists and as indicating the most probable steps in the processing of ribosomal precursor RNA.

It is a pleasure to acknowledge the capable assistance of Ricardo Fitten and Irene Fournier.

* This work was supported by grant CA-08416-02 from the National Institutes of Health and grant GB-5809 from the National Science Foundation.

[1] Perry, R. P., these PROCEEDINGS, **48**, 2179 (1962).
[2] Muramatsu, M., J. Hodnett, and H. Busch, *J. Biol. Chem.*, **241**, 1544 (1966).
[3] Chipchase, M. I. H., and M. L. Birnstiel, these PROCEEDINGS, **49**, 692 (1963).
[4] Edstrom, J. E., *J. Biophys. Biochem. Cytol.*, **11**, 549 (1961).
[5] Penman, S., I. Smith, and E. Holtzman, *Science*, **154**, 786 (1966).
[6] Penman, S., I. Smith, E. Holtzman, and H. Greenberg, *Natl. Cancer Inst. Monograph*, **23**, 489 (1966).
[7] Penman, S., in preparation.
[8] Laing, R., and S. Penman, in preparation.
[9] Scherrer, K., and J. E. Darnell, *Biophys. Biochem. Res. Commun.*, **9**, 451 (1962).
[10] Rake, A., and A. Graham, *Biophys. J.*, **4**, 267 (1964).
[11] Penman, S., *J. Mol. Biol.*, **17**, 117 (1966).
[12] Girard, M., H. Latham, S. Penman, and J. E. Darnell, *J. Mol. Biol.*, **11**, 187 (1965).
[13] Loening, U., *Biochem. J.*, **102**, 251 (1967).
[14] Eagle, H., *Science*, **130**, 432 (1959).
[15] Choules, G., and B. Zimm, *Analyt. Biochem.*, **13**, 336 (1965).
[16] Bray, G., *Analyt. Biochem.*, **1**, 279 (1960).
[17] Greenberg, H., and S. Penman, *J. Mol. Biol.*, **21**, 527 (1966).
[18] Zimmerman, E., and B. Holler, *J. Mol. Biol.*, **23**, 149 (1967).
[19] McCormick, W., and S. Penman, *Virology*, **31**, 135 (1967).
[20] Willems, M., and S. Penman, in preparation.
[21] Crowther, D., and J. Melnick, *Virology*, **15**, 65 (1961).
[22] Penman, S., and D. Summers, *Virology*, **27**, 614 (1965).
[23] Wagner, E., S. Penman, and V. Ingram, *J. Mol. Biol.*, in press.
[24] Studier, F. W., *J. Mol. Biol.*, **11**, 373 (1965).
[25] Vesco, C., and S. Penman, in preparation.
[26] Petermann, M. L., and A. Pavlovec, *Biochim. Biophys. Acta*, **114**, 264 (1966).
[27] Willems, M., E. Wagner, and S. Penman, in preparation.
[28] Penman, S., unpublished observations.

INDEX

200